SpringerBriefs in Physics

T0211127

For further volumes:
http://www.springer.com/series/8902

Mark A. Haidekker

Medical Imaging
Technology

Springer

Mark A. Haidekker
College of Engineering
University of Georgia
Athens, GA
USA

ISSN 2191-5423   ISSN 2191-5431 (electronic)
ISBN 978-1-4614-7072-4   ISBN 978-1-4614-7073-1 (eBook)
DOI 10.1007/978-1-4614-7073-1
Springer New York Heidelberg Dordrecht London

Library of Congress Control Number: 2013934273

Printed on acid-free paper

Springer is part of Springer Science+Business Media (www.springer.com)

Preface

Among many major developments in the medical field over the past two centuries, I personally consider three of them of particularly outstanding importance: The discovery of antibiotics in the late nineteenth century, the discovery of anesthesia during a period from the late eighteenth century to the mid-1800s, and the discovery of methods to look inside the human body without surgery. Out of these three, biomedical imaging is the youngest development, and its beginnings can be pinpointed to a very precise point in history—the discovery of the X-ray by C. W. Röntgen in 1895.

The history of medical imaging is a fascinating topic in itself, and it is briefly covered in Chap. 1. Most notably, the history of medical imaging is closely linked to the evolution of digital data processing and computer science, and to the evolution of digital electronics and the microprocessor. Medical imaging is truly interdisciplinary as it relies on physics, mathematics, biology, computer science, and engineering. This book tries to provide a solid foundation of the principles that lead to image formation. Specifically, many books on the same subject are intended for a medical or more general audience and treat the image formation process to some extent as a black box.

In this book, the image formation process can be followed from start to end, beginning with the question of how contrast is achieved. To this end, the source/detector systems that probe the tissue and provide the data necessary for image formation are explained. For each modality, this book explains the type of data collected, and how it is converted into an image. In addition, engineering aspects of the imaging devices, and a discussion of strengths and limitations of the modality can be found. In the first chapter, the basic concepts of contrast and resolution are introduced, and some concepts that are common to all modalities are explained. Subsequent chapters cover specific modalities.

We can broadly divide the imaging modalities into two groups, those with and those without the use of ionizing radiation. Unlike visible light, high-energy photons undergo only moderate scattering in tissue and can be used to penetrate the entire body. Most of the high-energy photons (X-ray and gamma radiation) follow a straight path, and certain geometrical assumptions are allowed that give rise to projection imaging (Chap. 2) and X-ray-based tomography (Chap. 3). Emission tomography, based on radioactive compounds that emit radiation from inside the body (Chap. 4) also belong to this category, since gamma radiation is used for the actual image formation. Imaging modalities with ionizing radiation share many common principles, and Chaps. 2 through 4 partly build on each other.

Magnetic resonance imaging and ultrasound imaging both use fundamentally different physical phenomena that are covered in Chaps. 5 and 6, respectively. Finally, Chap. 7 deals with recent developments both in the traditional modalities and in new modalities that are not yet widely used in clinical practice.

Since this book places considerable emphasis on the mathematical description of image formation, the reader should be familiar with calculus and have a basic understanding of differential equations. Although the Fourier transform is introduced in Sect. 1.4, some familiarity with the Fourier transform is helpful. Prior understanding of digital signal processing is helpful, too, although not a prerequisite for the topics covered in this book.

The chapters in this book can serve as an entry point for the in-depth study of individual modalities by providing the essential basics of each modality in a comprehensive and easy-to-understand manner. As such, this book is equally suitable as a textbook for undergraduate or graduate biomedical imaging classes and as a reference and self-study guide that prepare the reader for more specialized in-depth studies.

However, any one imaging modality could fill a whole book on its own, and for advanced study of a specific modality, more specialized books are also available. In-depth coverage of all modalities with an emphasis on the clinical application (but less emphasis on the mathematics) is provided in the book by Bushberg et al. [1]. A practice-oriented, user-friendly, yet highly detailed view of computed tomography can be found in [2], although it does not provide the mathematical details of CT image reconstruction. For readers who are interested in image reconstruction algorithms and their mathematical foundation, the books by Kak and Slaney [3] or by Herman [4] are recommended. MRI is a comprehensive subject, and in-depth coverage is provided in the work by Haake et al. [5]. A reference for ultrasound imaging can be found in the book by Hedrick et al. [6].

Among many individuals and colleagues who helped to shape this book with comments and discussions, I would particularly like to thank Erin E. Roberts for the help I received with some figures, Richard Speir and Dr. Adnan Mustafic for detailed revisions of the manuscript and helpful ideas how to improve the text, and to Prof. Qun Zhao and Prof. Geoff Dougherty for reviewing the book. Furthermore, I would like to express my gratitude toward the team at Springer: Christopher Coughlin and HoYing Fan, as well as the production team at SPS who ensured a smooth path from concept to publication.

Athens, January 2013   Mark A. Haidekker

Contents

Contents

Chapter 1
Introduction

Abstract "Medical imaging refers to several different technologies that are used to view the human body in order to diagnose, monitor, or treat medical conditions". All imaging modalities have in common that the medical condition becomes visible by some form of contrast, meaning that the feature of interest (such as a tumor) can be recognized in the image and examined by a trained radiologist. The image can be seen as a *model* of the imaged tissue. Images in the context of this book are digital. This implies a finite resolution with the pixel as the smallest element. Furthermore, all imaging modalities lead to some degradation of the image when compared to the original object. Primarily, the degradation consists of blur (loss of detail) and noise (unwanted contrast). Some underlying principles are common to all imaging modalities, such as the interpretation as a *system* and its mathematical treatment. The image itself can be seen as a multidimensional *signal*. In many cases, the steps in image formation can be seen as *linear systems*, which allow simplified mathematical treatment.

> *"Medical imaging refers to several different technologies that are used to view the human body in order to diagnose, monitor, or treat medical conditions. Each type of technology gives different information about the area of the body being studied or treated, related to possible disease, injury, or the effectiveness of medical treatment".*

This concise definition by the US Food and Drug Administration illuminates the goal of medical imaging: To make a specific condition or disease visible. In this context, *visible* implies that the area of interest is distinguishable in some fashion (for example, by a different shade or color) from the surrounding tissue and, ideally, from healthy, normal tissue. The difference in shade or color can be generalized with the term *contrast*.

The process of gathering data to create a visible model (i.e., the image) is common to all medical imaging technologies and can be explained with the simple example of a visible-light camera. The sample is probed with incident light, and reflected light carries the desired information. For example, a melanoma of the skin would reflect less light than the surrounding healthy skin. The camera lens collects some of the reflected light and—most importantly—focuses the light onto the film or

M. A. Haidekker, *Medical Imaging Technology*, SpringerBriefs in Physics, DOI: 10.1007/978-1-4614-7073-1_1, © The Author(s) 2013

image sensor in such a way that a spatial relationship exists between the origin of
the light ray and its location on the image sensor. The ability to spatially resolve
a signal (in this example, light intensity) is fundamental to every imaging method.
The ability to spatially resolve a signal can be fairly straightforward (for example,
following an X-ray beam along a straight path) or fairly complex (for example in
magnetic resonance imaging, where a radiofrequency signal is encoded spatially by
its frequency and its phase).

In the next step of the process, the spatially resolved data are accumulated. Once
again, the camera analogy is helpful. At the start of the exposure, the sensor array is
reset. Over the duration of the exposure, incoming light creates a number of electrical
charges that depends on the light intensity. At the end of the exposure, the charges
are transferred from the sensor to a storage medium. From here, the image would
typically be displayed in such a fashion that higher charge read-outs correspond to
higher screen intensity. In the camera example, the relationship between reflected
light intensity and displayed intensity is straightforward. In other cases, intensity
relates to different physical properties. Examples include X-ray absorption (which
gives X-ray images the characteristic negative appearance with bones appearing
bright and air dark), concentration of a radioactively labeled compound, or the time
it takes for a proton to regain its equilibrium orientation in a magnetic field.

The physical interpretation of image intensity is key to interpreting the image, and
the underlying physical process is fundamental to achieving the desired contrast. As
a consequence, the information encoded in the image varies fundamentally between
image modalities and, in some cases (such as MRI), even within the same modality.

The image is evaluated by an experienced professional, usually a radiologist. Even
in today's age of automated image analysis and computerized image understanding,
the radiologist combines the information encoded in the image with knowledge of
the patient's symptoms and history and with knowledge of anatomy and pathology to
finally form a diagnosis. Traditional viewing of film over a light box is still prominent,
even with purely digital imaging modalities, although more and more radiologists
make use of on-the-fly capabilities of the digital imaging workstation to view and
enhance images. Furthermore, computerized image processing can help enhance the
image, for example, by noise reduction, emphasizing edges, improving contrast, or
taking measurements.

1.1 A Brief Historical Overview

X-rays were discovered in 1895. Within less than a decade, which is an astonishingly
short time, X-ray imaging became a main-stream diagnostic procedure and was
adopted by most major hospitals in Europe and the USA. At that time, sensitivity
was low, and exposure times for a single image were very long. The biological
effects of X-rays were poorly explored, and radiation burns were common in the
early years of diagnostic—and recreational—X-ray use. As the pernicious effects of
ionizing radiation became better understood, efforts were made to shield operators

from radiation and to reduce patient exposure. However, for half a century, X-ray imaging did not change in any fundamental fashion, and X-ray imaging remained the only way to provide images from inside the body.

The development of sonar (sound navigation and ranging) eventually led to the next major discovery in biomedical imaging: ultrasound imaging. After World War II, efforts were made, in part with surplus military equipment, to use sound wave transmission and sound echoes to probe organs inside the human body. Ultrasound imaging is unique in that image formation can take place with purely analog circuits. As such, ultrasound imaging was feasible with state-of-the-art electronics in the 1940s and 1950s (meaning: analog signal processing with vacuum tubes). Progress in medical imaging modalities accelerated dramatically with the advent of digital electronics and, most notably, digital computers for data processing. In fact, with the exception of film-based radiography, all modern modalities rely on computers for image formation. Even ultrasound imaging now involves digital filtering and computer-based image enhancement.

In 1972, Geoffrey Hounsfield introduced a revolutionary new device that was capable of providing cross-sectional, rather than planar, images with X-rays. He called the method *tomography*, from the Greek words to cut and to write [7]. The imaging modality is known as computed tomography (CT) or computer-aided tomography (CAT), and it was the first imaging modality that *required* the use of digital computers for image formation. CT technology aided the development of emission tomography, and the first CT scanner was soon followed by the first positron emission tomography scanner.

The next milestone, magnetic resonance imaging (MRI), was introduced in the late 1970s. MRI, too, relies on digital data processing, in part because it uses the Fourier transform to provide the cross-sectional image. Since then, progress became more incremental, with substantial advances in image quality and acquisition speed. The resolution and tissue discrimination of both CT and MRI, for example, that today's devices are capable of, was literally unthinkable at the time these devices were introduced. In parallel, digital image processing and the digital imaging workstation provided the radiologist with new tools to examine images and provide a diagnosis. Three-dimensional image display, multi-modality image matching, and preoperative surgery planning were made possible by computerized image processing and display.

A present trend exists toward the development of imaging modalities based on visible or infrared light. Optical coherence tomography (OCT) became widely known in the 1990s and has evolved into a mainstream method to provide cross-sectional scans of the retina and skin. Other evolving optical modalities, such as diffuse optical tomography, have not reached the maturity level that would allow its use in medical practice.

1.2 Image Resolution and Contrast

Digital images are discretely sampled on a rectangular grid. A digital camera again illustrates the nature of a digital image: the camera sensor is composed of

Fig. 1.1 Sketch of a magnified part of a digital camera image sensor. Each sensor cell consists of a light-sensitive photodiode (gray-shaded area) and associated amplifier and driver circuitry (hatched region). Each sensor cell averages the light across its sensitive surface and provides one single intensity value

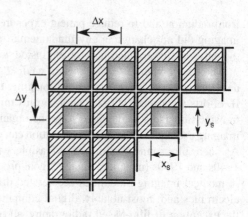

millions of light-sensitive cells. A sketch of a few cells, strongly magnified, is shown in Fig. 1.1. Each single sensor cell is composed of a light-sensitive semiconductor element (photodiode) and its associated amplifier and drive circuitry. The cells are spaced Δx apart in the horizontal direction, and Δy in the vertical direction. The actual light-sensitive area is smaller, x_s by y_s. To illustrate these dimensions, let us assume a 12-megapixel sensor with 4,000 cells in the horizontal and 3,000 cells in the vertical direction. When the overall dimensions of the sensor chip are 24 by 18 mm, we know $\Delta x = \Delta y = 6\,\mu$m. Depending on the chip design, the photodiode occupies most of the space, for example, $x_s = y_s = 5\,\mu$m. Irrespective of the amount of detail in the image projected onto the sensor, detail information smaller than the size of a sensor cell is lost, because the photodiode averages the intensity over its surface, and the surrounding driver is not sensitive to light. Each cell (i.e., each pixel), therefore, provides one single intensity value that is representative of the area it occupies.

The spatial resolution of the most important medical imaging modalities spans a wide range. Planar X-ray imaging can achieve a spatial resolution of up to 10 μm, in part limited by the film grain. Digital X-ray sensors can achieve a similarly high resolution, although 20–50 μm pixel size is more common. With CT, in-plane pixel sizes between 0.1 and 0.5 mm are common in whole-body scanners. MRI scanners have typical in-plane pixels of 0.5–1 mm. Due to the different detector system, radionuclide imaging modalities (SPECT and PET) have pixel sizes in the centimeter range. Ultrasound resolution lies between CT and MRI.

The sensor is not the only limiting factor for the spatial resolution. An ideally focused light source is spread out by the camera lens, primarily as a consequence of lens shape approximations and light diffraction. The image of a point source is called the *point-spread function*. The importance of the point-spread function is demonstrated in Fig. 1.2. The image shows photos of tightly focused laser beams taken with a digital SLR camera from 2 m distance. It can be seen that the image of a single beam shows a Gaussian profile (Fig. 1.2a). An ideal imaging apparatus would provide a delta function (i.e., a cylinder of one pixel width). The point-spread function can be quantified by its full width at half-maximum (FWHM), that is, the

Fig. 1.2 Point-spread function of a digital camera, shown in grayscale representation and as an elevation map where intensity translates into height. The pixel size Δx and Δy is indicated. **a** The image of a highly focused laser beam has a Gaussian shape in the image. The full width at half-maximum (FWHM) spread is 6 pixels. **b** Two closely spaced sources can be distinguished if their distance is larger than the FWHM. **c** Two sources that are more closely spaced than the FWHM become indistinguishable from a single source

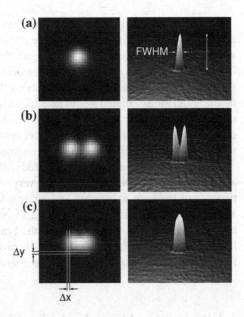

width of the point image where it drops to one half of its peak value (Fig. 1.2a). In this example, we observe a FWHM of 6 pixels. As long as two closely spaced point sources are further apart than the FWHM, they can be distinguished as two separate peaks (Fig. 1.2b), which is no longer possible when the point sources are closer than the FWHM (Fig. 1.2c).

Clearly, the point-spread function poses a limit on the spatial resolution, often more so than the detector size. In X-ray imaging, for example, one factor that determines the point-spread function is the active area of the X-ray tube. In ultrasound imaging, factors are the length of the initial ultrasound pulse and the diameter of the ultrasound beam. Furthermore, the wavelength of the sound wave itself is a limiting factor.

The image values are stored digitally. A certain number of bits is set aside for each cell (each pixel). Since each bit can hold two values (one and zero), a n-bit pixel can hold 2^n discrete intensity levels. Color photos are commonly stored with 24 bits per pixel, with 8 bits each for the three fundamental colors, red, green, and blue. For each color, 256 intensity levels are possible. Most magnetic resonance and ultrasound images are also stored with 8 bits depth, whereas computed tomography normally provides 12 bits.

The pixel size determines the absolute limit for the spatial resolution, and the bit depth determines the contrast limit. Consider an 8-bit image: the intensity increase from one discrete image value to the next is 0.39 % of the maximum value. Any smaller intensity variations cannot be represented. The error that is associated with rounding of a continuous signal to the next possible image value is referred to as *digitization noise*.

   Noise is introduced in several steps of the acquisition and processing chain. Both
the sensors and the amplifiers introduce noise components, particularly when weak
signals need to be amplified by a large gain factor. Examples are the RF echo signal
in MRI and the ultrasound echo in ultrasound imaging. To some extent noise can be
suppressed with suitable filters, but the side-effect is a broadening of the point-spread
function and the associated loss of detail. Conversely, any filter that tries to counteract
the point-spread function increases the noise component. The noise component is
critical for the overall image quality, because noise can "bury" detail information
from small objects or objects with low contrast.

   The ability to provide a specific, desired contrast depends strongly on the modality.
X-ray imaging, for example, provides very strong contrast between bone and soft
tissue, and between soft tissue and air (e.g., in images of the lung or the chest).
Magnetic resonance imaging shows high contrast between different types of soft
tissue (e.g., gray and white matter of the brain), but bone and air are dark due to
the absence of water. Ultrasound generally provides good tissue contrast, but suffers
from a high noise component, visible as characteristic ultrasound speckles.

1.3 Systems and Signals: A Short Introduction

System is a broad term that encompasses any assembly of interconnected and interacting components that have measurable behavior and a defined response to a defined
manipulation of its parts. Any device that provides a medical image is a system in
this definition, and it consists in turn of several components that can be seen as systems themselves. Systems have inputs and outputs. One example for a system is an
X-ray detector. The number of X-ray photons hitting the conversion layer can be
interpreted as the system input. The detector provides a voltage that is proportional
to the incident photon flux, and this voltage is the output. Similarly, a computer algorithm for image reconstruction is a system. In a computed tomography scanner, for
example, the input to the image reconstruction algorithm is the X-ray intensity as
a function of the scan angle and position, and the output is a two-dimensional map
(i.e., cross-section) of X-ray absorption coefficients.

   The input and output to a system can be interpreted as *signals*. Often, a signal is
understood as a function of time, but in imaging devices, signals are functions of a
spatial coordinate. In the most general form, imaging devices process signals of the
form $f(x, y, z, t)$, that is, a quantity that depends on a location in (three-dimensional)
space and on time. Often, simplifications can be made when a signal is approximately
constant over the image acquisition time, or when a signal is obtained only within one
plane. An example is shown in Fig. 1.3, where components of the scanning and image
reconstruction process are shown as blocks with signals represented by arrows.

   In any system, the output signal can be described mathematically for a given input
signal. The X-ray detector, for example, converts X-ray photon flux $I(t, \theta)$ into a
proportional voltage $U(t, \theta)$:

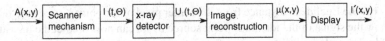

Fig. 1.3 Systems interpretation of a computed tomography scanner. Components of the system (in itself systems) are represented by blocks, and signals represented by arrows. The original object has some property A, for example, X-ray absorption that varies within the x, y-plane. The detector collects X-ray intensity $I(t, \theta)$ as a function of scan direction t and scan angle θ in the (x, y)-plane, and provides a proportional voltage $U(t, \theta)$. In the image formation stage, these data are transformed into a cross-sectional map of apparent X-ray opaqueness, $\mu(x, y)$. Finally, the display outputs a light intensity $I'(x, y)$ that is proportional to $\mu(x, y)$ and approximates $A(x, y)$

$$U(t, \theta) = \alpha \cdot I(t, \theta) \tag{1.1}$$

where α is the gain of the X-ray detector. Similarly, the image reconstruction stage approximates the inverse Radon transform $\mathscr{R}^{-1}$ (see Sect. 3.1.1):

$$\mu(x, y) = \mathscr{R}^{-1}\{U(t, \theta)\}. \tag{1.2}$$

A special group of systems are *linear, time-invariant systems*. These systems are characterized by three properties,

- Linearity: If y is the output for a given input x, then a change of the magnitude of the input signal by a constant factor a (i.e., we input ax) leads to a proportional output signal ay.
- Superposition: If a system responds to an input signal x_1 with the output signal y_1 and to a different input signal x_2 with y_2, then the sum of the input signals $x_1 + x_2$ will elicit the response $y_1 + y_2$.
- Time-invariance: If $y(t)$ is the time-dependent output signal for a given input signal $x(t)$, then the application of the delayed signal $x(t - \tau)$ causes an identical, but equally delayed response $y(t - \tau)$. In images, time-invariance translates into shift-invariance. This means that an operator that produces an image $I(x, y)$ from an input image produces the same image, but shifted by Δx, Δy, when the input image is shifted by the same distance.

Figure 1.3 provides a different view of the point-spread function: we can see that an object (the original tissue property $A(x, y)$) is probed by some physical means. The image formation process leads to the display of an image $I'(x, y)$, which differs from $A(x, y)$. Referring back to Fig. 1.2, we can see that the image functions of the laser dots are superimposed (i.e., added together). With the superposition principle, we can examine each individual pixel separately and subject it to the point-spread function, then add the results. Very often, the point-spread function has Gaussian character, and we can model the peak seen in Fig. 1.2a as

$$g(r) = \frac{1}{\sigma\sqrt{2\pi}} \exp\left(-\frac{r^2}{2\sigma^2}\right) \tag{1.3}$$

where r is the Euclidean distance from the center pixel of the point source (x_0, y_0). If we know the signal of the idealized point source S, we can now predict the measured (i.e., blurred with the PSF) intensity for each pixel (x, y):

$$I(x, y) = \frac{S}{\sigma\sqrt{2\pi}} \exp\left(-\frac{(x - x_0)^2 + (y - y_0)^2}{2\sigma^2}\right) = S \cdot g(x - x_0, y - y_0) \quad (1.4)$$

where $g(x - x_0, y - y_0)$ should be seen as a generalized point-spread function whose center is shifted to the center of the point source. Consequently, we can express the image formed by two point sources of strength S_0 and S_1 and centered on (x_0, y_0) and (x_1, y_1), respectively, as the superposition of the image functions

$$I(x, y) = S_0 \cdot g(x - x_0, y - y_0) + S_1 \cdot g(x - x_1, y - y_1). \quad (1.5)$$

This concept can be further generalized. Assume that we have an idealized (but inaccessible) source image $S(x, y)$ and we measure the image $I(x, y)$ with an imaging device that makes it subject to the point-spread function g. In this case, we can subject *each individual pixel* of $S(x, y)$ to the point-spread function and recombine them by addition:

$$I(x', y') = \sum_y \sum_x S(x, y) \cdot g(x' - x, y' - y). \quad (1.6)$$

The sum in Eq. 1.6 needs to be evaluated for all pixels (x', y') of the target image I. Equation 1.6 describes the two-dimensional discrete *convolution* of the source image S with a convolution function (often called *convolution kernel*) g. Since any bright pixel spreads out and influences its neighbors (thus, point *spread* function), sharp transitions are softened, and detail is lost. The effect is demonstrated in Fig. 1.4, where the idealized image $S(x, y)$ (Fig. 1.4a) has been subjected to a simulated point-spread function, in this case, a Gaussian function with $\sigma \approx 3.5$

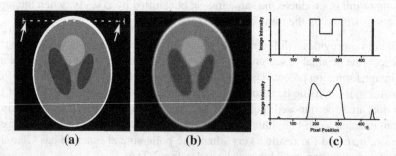

(a) (b) (c)

Fig. 1.4 Illustration of the effects of a Gaussian point-spread function. **a** Idealized image (note the two added white dots in the top left and top right corners indicated by arrows). **b** Image obtained after a process with a Gaussian point-spread function. **c** Intensity profiles along the horizontal dashed line in **a**. It can be seen that sharp transitions are softened, because higher image values also influence their neighbors. Point sources assume a Gaussian-shaped profile

pixels to reveal the actual image $I(x, y)$ (Fig. 1.4b). The line profiles (Fig. 1.4c) help illustrate how sharp transitions are blurred and how isolated points assume a Gaussian shape.

1.4 The Fourier Transform

The Fourier transform is one of the most important linear operations in image processing, and it is fundamental to most imaging modalities. Intuitively, a *transform* reveals a different aspect of the data. In the case of the Fourier transform, it shows the distribution of harmonic content—how the signal is composed of periodic oscillations of different frequency and amplitude. For example, a time-dependent oscillation $f_\omega(t) = A \sin(\omega t)$ could be described by its amplitude A and its frequency ω. In a diagram $f_\omega(t)$ over t, we obtain an oscillation. In a diagram of amplitude over frequency, the same signal is defined by a single point at (A, ω). Superimposed sine waves would be represented by multiple points in the diagram of amplitude over frequency. In this simplified explanation, a phase shift $f(t) = A \sin(\omega t + \varphi)$ cannot be considered, because a third dimension becomes necessary to include A, ω, and φ. The Fourier transform uses sine and cosine functions to include the phase shift, and each harmonic oscillation becomes $f(t) = a \cos(\omega t) + b \sin(\omega t)$.

Fourier's theorem states that *any periodic signal $s(t)$ can be represented as an infinite sum of harmonic oscillations*, and the Fourier synthesis of the signal $s(t)$ can be written as

$$s(t) = \frac{a_0}{2} + \sum_{k=1}^{\infty} a_k \cdot \cos(kt) + b_k \cdot \sin(kt) \tag{1.7}$$

where a_k and b_k are the Fourier coefficients that determine the contribution of the kth harmonic to the signal $s(t)$. For any given signal $s(t)$, the Fourier coefficients can be obtained by *Fourier analysis*,

$$a_k = \frac{1}{\pi} \int_{-\pi}^{\pi} s(t) \cos(kt) dt$$

$$b_k = \frac{1}{\pi} \int_{-\pi}^{\pi} s(t) \sin(kt) dt. \tag{1.8}$$

Equation 1.7 describes the synthesis of a signal from harmonics with integer multiples of its fundamental frequency. The spectrum (i.e., the sequence of a_k and b_k) is discrete. The continuous Fourier transform is better derived from a different form of Fourier synthesis that uses a continuous spectrum $a(\omega)$ and $b(\omega)$:

$$s(t) = \int_{\omega} a(\omega) \cdot \cos(\omega t) + b(\omega) \cdot \sin(\omega t) d\omega. \tag{1.9}$$

The integration takes place over all possible frequencies ω. Since the basis functions (sin and cos) in Eq. 1.9 are orthogonal, we can express the Fourier synthesis in terms of a complex harmonic oscillation $e^{j\varphi} = \cos(\varphi) + j\sin(\varphi)$. Fourier synthesis restores a signal from its spectrum and corresponds to the inverse Fourier transform $\mathscr{F}^{-1}$, whereas the Fourier analysis, which provides the spectrum of a signal, is referred to as the actual Fourier transform $\mathscr{F}$:

$$S(\omega) = \mathscr{F}\{s(t)\} = \int_{-\infty}^{\infty} s(t)\exp(-j\omega t)dt$$

$$s(t) = \mathscr{F}^{-1}\{S(\omega)\} = \frac{1}{2\pi} \int_{-\infty}^{\infty} S(\omega)\exp(j\omega t)d\omega. \tag{1.10}$$

Equation 1.10 defines the Fourier transform in terms of the angular frequency $\omega = 2\pi f$. In some cases, it is more convenient to express the spectrum $S(f)$ as a function of the linear frequency f, for which the Fourier transform becomes

$$S(f) = \mathscr{F}\{s(t)\} \int_{-\infty}^{\infty} s(t)\exp(-2\pi jft)dt$$

$$s(t) = \mathscr{F}^{-1}\{S(f)\} = \int_{-\infty}^{\infty} S(f)\exp(2\pi jft)d\omega. \tag{1.11}$$

To explain the significance of the Fourier transform, let us consider two examples. First, in magnetic resonance imaging, we deal with signals that are caused by protons spinning at different speeds (cf. Sect. 5.4.3). The angular frequency of the protons increases along one spatial axis (let us call it the y-axis), and the protons emit a signal whose strength is determined, among other factors, by the number of protons at any point along the y-axis. The signal can be collected by an antenna, but the antenna only provides the additive mix of all signals. We can, however, obtain the local proton density by using the relationship $\omega = \omega_0 + m \cdot y$, where m is the rate of change of the frequency along the y-axis. The antenna provides a signal $s(t)$, which we subject to the Fourier transform. The resulting harmonic content $S(\omega) = S(\omega_0 + m \cdot y)$ is directly related to the signal strength at any point along the y-axis and therefore to the proton density.

Second, it is sometimes desirable to have a signal that contains all frequencies in a limited range (i.e., a broadband signal). We can ask the question, how would a broadband signal $b(t)$ look like for which the spectral component is unity for all frequencies between $-f_0$ and $+f_0$?[1] To answer this question, we use the inverse

---

[1] Examining negative frequencies is not unreasonable. Equation 1.9 holds for $\omega < 0$, and the Fourier transform shows some symmetry. The Fourier transform has a number of very interesting properties, but they go beyond the scope of this book.

Fourier transform (Eq. 1.11) with the description of the broadband signal,

$$B(f) = \begin{cases} 1 & \text{for } -f_0 < f < f_0 \\ 0 & \text{otherwise} \end{cases} \tag{1.12}$$

which leads to the following integral where the limits of the bandwidth determine the integration bounds:

$$b(t) = \int_{-f_0}^{f_0} e^{2\pi jft} \mathrm{d}f = \frac{1}{2\pi jt} \left[e^{2\pi jf_0 t} - e^{-2\pi jf_0 t} \right]. \tag{1.13}$$

Fortunately, Euler's relationship allows us to simplify the expression in square brackets to $2j \sin(2\pi f_0 t)$, and the imaginary unit j cancels out. We therefore obtain our broadband signal as

$$b(t) = \frac{\sin(2\pi f_0 t)}{\pi t}. \tag{1.14}$$

For $f_0 = 1/2$, Eq. 1.14 describes the well-known *sinc* function, and it can be shown that the boxcar function (Eq. 1.12) and the sinc-function are a Fourier transform pair, meaning, a square pulse in the time domain has a sinc-like spectrum, and a sinc-like function has a boxcar-type spectrum.

Since digital signals and digital images are discretely sampled, we need to take a look at the *discrete Fourier transform*. In the one-dimensional case, the signal $s(t)$ exists as a set of N discretely sampled values s_k, obtained at $t = k\Delta t$. Here, Δt is the sampling period. In the discrete world, the integral corresponds to a summation, and the discrete Fourier transform becomes

$$S_u = \mathscr{F}\{s_k\} = \frac{1}{N} \sum_{k=0}^{N-1} s_k \exp\left(-2\pi j \frac{u \cdot k}{N} \right) \tag{1.15}$$

where u is the discrete frequency variable, and the sum needs to be evaluated for $0 \le u \le N/2$. Equation 1.15 does not consider the sampling rate, and Δt needs to be known to relate u to any real-world units. Any spectral component S_u has the corresponding frequency f_u,

$$f_u = \frac{u}{N \cdot \Delta t}. \tag{1.16}$$

Note that Eq. 1.16 is not limited to sampling in time. When Δt is a time interval, f_u has units of frequency (i.e., inverse seconds). However, a signal can be sampled with discrete detectors along a spatial axis (see, for example, Fig. 1.1). In this case, the sampling interval has units of distance, and f_u has units of inverse distance. This is referred to as *spatial frequency*. An example to illustrate spatial frequency is a diffraction grating, which causes interference patterns with a certain spatial distance.

For example, if an interference maximum occurs every 0.2 mm, the corresponding spatial frequency is $5\,\text{mm}^{-1}$ (or 5 maxima per mm).

We can see from Eq. 1.15 that choosing $u = N$ yields the same result as $u = 0$. For increasing u, therefore, the spectrum repeats itself. Even more, the symmetry of the complex exponential in Eq. 1.15 provides us with $S^*_{-u} = S_u$, where S^* indicates the conjugate-complex of S. For this reason, we gain no new information from computing the discrete Fourier transform for $u > N/2$. By looking at Eq. 1.16, we can see that the frequency at $u = N/2$ is exactly one half of the sampling frequency. This is the maximum frequency that can be unambiguously reconstructed in a discretely-sampled signal (known as the Shannon sampling theorem). The frequency $f_N = 1/2\Delta t$ is known as the *Nyquist frequency*. In the context of Fig. 1.1, we briefly touched on the loss of detail smaller than the sensor area. Here, we have approached the same phenomenon from the mathematical perspective. The situation becomes worse if the signal actually contains frequency components higher than the Nyquist frequency, because those spectral components are *reflected* into the frequency band below f_N, a phenomenon known as *aliasing*. Aliasing is not limited to signals sampled in time. Spatial discretization of components with a higher spatial frequency than $1/2\Delta x$ leads to Moiré patterns.

For images (i.e., discretely-sampled functions in two dimensions), the Fourier transform can be extended into two dimensions as well. Because of the linearity of the Fourier transform, we can perform the row-by-row Fourier transform in one dimension and subject the result to a column-by-column Fourier transform in the orthogonal dimension:

$$S(u, v) = \mathscr{F}\{s(x, y)\} = \frac{1}{MN} \sum_{x=0}^{N-1} \sum_{y=0}^{M-1} s(x, y) \exp\left(-2\pi j \left[\frac{ux}{N} + \frac{vy}{M}\right]\right). \quad (1.17)$$

The Fourier transform now has two orthogonal frequency axes, u and v. The inverse Fourier transform is

$$s(x, y) = \mathscr{F}^{-1}\{S(u, v)\} = \sum_{u=0}^{N-1} \sum_{v=0}^{M-1} S(u, v) \exp\left(2\pi j \left[\frac{ux}{N} + \frac{vy}{M}\right]\right). \quad (1.18)$$

The two-dimensional inverse Fourier transform finds its application in the reconstruction process in computed tomography and magnetic resonance. In both cases, one-dimensional Fourier-encoded information is gathered and used to fill a 2D Fourier-domain spaceholder. Once the spaceholder is completely filled, the inverse Fourier transform yields the cross-sectional reconstructed image.

Chapter 2
X-Ray Projection Imaging

Abstract X-ray imaging is the oldest medical imaging modality, which found its way into medical practice shortly after the discovery of the X-rays in 1895. X-ray imaging is a projection technique, and image formation takes place traditionally on photosensitive film, although direct digital X-ray imaging is becoming more and more common. In its most common form, X-ray imaging is a qualitative modality. X-rays are high-energy photons, and atomic interaction with inner shell electrons is fundamental to both X-ray production and generation of X-ray contrast. Soft-tissue contrast is comparatively low, but bone and air provide excellent contrast. In some cases, contrast can be enhanced with contrast agents. An undesirable (but unavoidable) side-effect of the photon-atom interaction is the ionization of tissue along the beam path, which can lead to radiation damage. X-ray images can reveal very subtle features, and its popularity is further enhanced by the relatively inexpensive equipment and the straightforward imaging procedure.

2.1 X-Ray Generation

X-rays are generated when kinetic electrons interact with a solid target and undergo sudden deceleration ("braking radiation" or "bremsstrahlung"). As electrons lose some of their kinetic energy, this energy is released as an X-ray photon. In an X-ray tube, electrons are emitted into the vacuum at the cathode and accelerated toward the anode in a high-voltage electrostatic field. In the dense metallic anode material, the electrons lose part of their energy when they are deflected in the Coulombic field of the nucleus (Fig. 2.1). The extreme case is a direct collision of a high-energy electron with a nucleus, whereby the electron loses all of its kinetic energy, and a photon of the same energy is emitted. If the electron comes in close proximity of the nucleus, it is deflected by electrostatic forces. In the process, the electron loses some (but not all) of its energy, and a photon is emitted that carries this amount of energy. The greater the distance to the nucleus, the lower the energy loss of the electron and the lower the photon energy. Electrons can undergo multiple interactions with

M. A. Haidekker, *Medical Imaging Technology*, SpringerBriefs in Physics,
DOI: 10.1007/978-1-4614-7073-1_2, © The Author(s) 2013

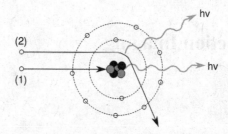

Fig. 2.1 X-ray generation by electron deceleration. High-energy electrons come in proximity of a nucleus of the anode material and are deflected in the Coulombic field of the nucleus. In the extreme case (*1*), the electron collides directly with the nucleus, thereby releasing all of its kinetic energy into one X-ray photon. A close "fly-by" (*2*) is much more probable, but the electron loses less energy, and the resulting X-ray photon has a lower energy and a correspondingly longer wavelength

nuclei before they expend all of their energy. Since the kinetic energy E_{kin} of the electrons only depends on the electrostatic potential (i.e., the anode voltage U), the maximum photon energy E_{max} and its corresponding shortest wavelength λ_{min} can be computed as

$$E_{kin} = \frac{1}{2}m_e v^2 = e \cdot U \tag{2.1}$$

$$E_{kin} = E_{max} = \frac{hc}{\lambda_{min}} \tag{2.2}$$

where m_e is the electron mass, e its charge, v its velocity upon impact with the anode, h is Planck's constant, and c the speed of light.

The X-ray radiation generated in a regular X-ray tube has therefore a continuous energy *spectrum*. A direct collision with the nucleus is rare, and photons with the highest possible energy occur in the spectrum with a low probability. The probability increases with increasing electron-nucleus distance and correspondingly lower photon energy.

Interactions of the high-energy electrons with shell electrons of the anode material play a significant role. If high-energy electrons collide with shell electrons, the atom of the anode material can be excited or ionized, usually by ejection of a k-shell electron. As the k-shell vacancy gets filled, energy is released—again as X-ray radiation. Since the binding energies of the target material are fixed, the X-ray energy that is emitted from shell electrons has a very narrow energy spectrum (Fig. 2.2). These narrow spikes in the energy spectrum are referred to as *characteristic X-rays*, because they are characteristic for the target material.

Example: In a tube with a tungsten anode, the anode voltage is 75 kV. The highest possible energy is 75 keV (1 eV = 1.602×10^{-19} J). The shortest wavelength $\lambda_{min} = 16.5$ pm (pm = picometers = 10^{-12} m). In addition, the binding energy of the k-shell electrons of tungsten is 69.5 keV, and the binding energy of the l-shell electrons is 12.1 keV. If an electron is ejected from the k-shell, and the vacancy is filled from the

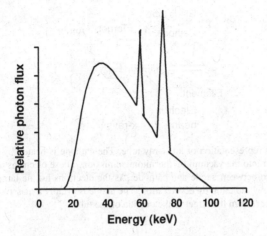

Fig. 2.2 Typical X-ray spectrum of a tube with a tungsten anode operated at 100 kV. The maximum possible photon energy is 100 keV, but lower energies are more probable. Very low energies (25 keV and below) get absorbed by the anode material and by the glass wall of the tube, and do not occur in the spectrum. Two prominent peaks of characteristic X-rays can be seen at 69.5 keV (k-shell) and 57.4 keV (k−l transition)

l-shell, the energy difference of 57.4 keV is emitted as an X-ray photon. Note that the binding energy of higher shells is too low to cause X-ray radiation. Note also that the characteristic X-ray radiation can only occur if the kinetic energy of the incident electrons is high enough to allow at least the k−l transition. For example, if the tube is operated at 50 kV, no characteristic X-rays from tungsten can be seen.

For comparison, visible light (green) has a wavelength of 520 nm and an associated energy of 2.4 eV. The lowest energy to affect the atomic shell is the k−l excitation of hydrogen and requires 10.1 eV with an associated wavelength of 123 nm. Diagnostic X-rays lie approximately in the range from 30 to 120 keV with wavelengths in the picometer range.

2.1.1 The X-Ray Tube

The actual X-ray tube consists of a vacuum glass cylinder with the anode and cathode inside. Electrical contacts allow connecting of the cathode, its heating filament, and the anode to the power sources. The anode is usually made of copper for good heat conduction with a small target area of tungsten or rhodium. Tungsten melts at about 3700 K, and copper has a melting point of about 1360 K. Heat dissipation is important, because only a small amount of the electron beam's kinetic energy is actually emitted as X-rays. X-ray tubes have a fairly poor degree of efficiency η, which can be approximated by

$$\eta = k \cdot U \cdot Z \tag{2.3}$$

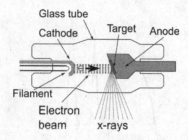

Fig. 2.3 Schematic representation of an X-ray tube. The cathode is heated by the filament, and electrons are ejected into the vacuum by thermionic emission. These electrons are accelerated by the electrostatic field between anode and cathode. As the electrons hit the target material of the anode, they emit X-ray radiation by deceleration. The anode is usually a massive body of metal to facilitate heat transport from the target to the outside of the tube

where k is a proportionality constant with approximately $k = 1.1 \times 10^{-9}$, U is the anode voltage, and Z is the atomic number of the anode material, for example, $Z = 74$ for tungsten. Typically, about 1 % of the energy is emitted as X-rays, and 99 % is lost as heat. Assuming a 70 kV tube with a tube current of 5 mA, total energy dissipation is 350 W, comparable to a hair dryer on a medium setting. For this reason, large X-ray tubes designed for continuous operation use rotating anodes: the anode is a large disc, driven by a motor. The disc is beveled, and electrons hit only a small active region on the beveled rim. The rotating anode shows much lower local heating than a fixed anode.

The cathode is heated by a filament, and electrons are injected into the vacuum by thermionic emission. Electrons from the cathode are accelerated toward the anode, gathering kinetic energy in the process. When the electrons hit the target material of the anode, they lose their kinetic energy, and X-rays are emitted.

A simplified sketch of an X-ray tube is shown in Fig. 2.3. The anode target is usually angled to direct a large part of the X-ray radiation perpendicular to the electron beam.

The filament current allows to control the cathode temperature. The thermionic effect is highly temperature dependent, and the current flux S (current per area unit of emitting cathode material) is described by the *Richardson equation*,

$$S = A_R \cdot T^2 \cdot \exp\left(-\frac{E_A}{kT}\right) \tag{2.4}$$

where A_R is the Richardson constant, E_A is the activation energy (also called work function), k is Boltzmann's constant, and T is the absolute temperature. A_R and E_A are material constants of the cathode material. To achieve a reasonable tube current, a cathode temperature between 1800 and 2000 K is desirable (Fig. 2.4). Special cathode materials, such as cesium on tungsten or barium oxide, have a lower activation energy and require lower temperatures for the desired tube current. The activation energy

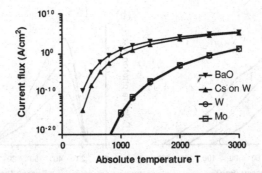

Fig. 2.4 Temperature-dependency of the cathode current flux by thermionic emission (Eq. 2.4). Typical tube currents are in the milliampere range, and a tungsten cathode needs to be heated to near its melting point. Some special materials, such as cesium or barium oxide, have a lower activation energy and require lower temperatures

Table 2.1 Richardson constant A_R and activation energy E_A of some cathode materials

Material	E_A (eV)	A_A (A / cm^2 K^2)
Tungsten	4.53	60
Molybdenum	4.43	55
Cs on W	1.36	3.2
BaO	0.99	1.18

E_A and Richardson constant A_R for some common cathode materials is listed in Table 2.1.

The X-ray tube has two main "knobs" which influence the emitted radiation (see also Fig. 2.5):

1. The anode voltage: Increasing the anode voltage increases the maximum energy of the X-ray photons, and it also increases the total number of photons emitted. An empirical relationship between the anode voltage U and the number of photons emitted N is

$$N = k \cdot Z \cdot U^m \tag{2.5}$$

   where k is a tube-dependent proportionality constant, Z is the atomic number of the target material, and m is a material-dependent exponent, typically $m \approx 2$.

2. The tube current: By modifying the filament current and with it the cathode temperature, the tube current can be controlled. Increasing the tube current by increasing the filament temperature increases X-ray photon flux, but does not change the highest energy of the emitted photons.

In practice, the X-ray tube is housed in a lead-shielded assembly to prevent X-ray radiation in any direction except the aperture of the X-ray housing. Often, collimators (movable lead blades) allow adjustment of the aperture. With a suitably positioned

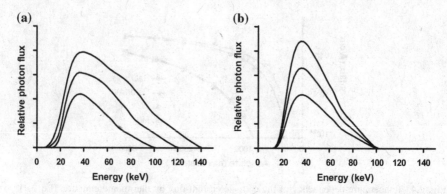

Fig. 2.5 Qualitative changes of the emission spectra (characteristic X-rays omitted!) of an X-ray tube when the anode voltage is increased (**a**) and when the filament current is increased (**b**). An increase of the anode voltage (in this example from 100 to 120 and 140 kV) increases the maximum photon energy, but indirectly also increases the tube current and therefore the total photon flux. Conversely, an increase of the filament current (in this example by 20 and 50 %) increases the total photon flux, but not their energy distribution. Notably, the maximum energy remains the same

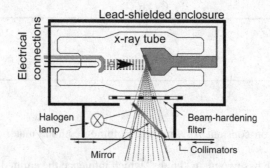

Fig. 2.6 Schematic representation of an X-ray tube assembly with its enclosure and additional features. A beam-hardening filter removes low-energy X-rays from the beam. Subsequently, movable collimators allow control of the illuminated area. A visible-light lamp and a mirror provide visual information on the illuminated area. The entire enclosure is heavily lead-shielded to minimize radiation everywhere except at the target area of the patient

lamp and mirror, the area that will be exposed to X-rays can be highlighted beforehand with visible light. Lastly, a beam-hardening filter (sometimes removable) absorbs some of the low-energy radiation and reduces the patient dose. Figure 2.6 shows a sketch of a complete X-ray tube assembly with its housing, beam-hardening filter, collimators, and guide lamp.

A complete X-ray generator contains several more components. The key element is the high-voltage generator, built around a transformer and frequently a high-voltage cascade, such as a Cockroft-Walton multiplier. Through a resistor voltage divider, the anode voltage can be measured. For example, if the voltage divider has a ratio of 1:10,000, an anode voltage of 150 kV is converted down to 15 V, which is in the voltage range of conventional operational amplifiers. A closed-loop feedback control system is therefore feasible that tightly controls the anode voltage. Similarly, a current shunt resistor allows to measure the tube current and, through a feedback control system, to control the filament current. In addition, the total dose emitted by the X-ray tube can be measured and used to shut off the high voltage after the exposure is finished. Selection of the exposure time adds a third "knob", but tube current and exposure time have similar effects.

2.1.2 A Focus on Geometry

The anode is usually angled to direct a large part of the X-ray radiation perpendicular to the electron beam. Moreover, the anode angle allows to balance illumination strength of the tube with its ability to resolve small details. The cathode emits an electron beam of diameter d (Fig. 2.7). The projected focal spot, as seen from the detector, has a diameter $D = d/\tan\alpha$. Both the anode angle α and the electron beam diameter d are design parameters:

- A thin electron beam with a small diameter d has a small focal spot with good detail resolution, but poor overall intensity, because the maximum tube current is limited.
- An electron beam with a large diameter d and a steep angle $\alpha \to 90°$ also has a small focal spot, but larger tube currents are possible. Because of the large angle, however, the illuminated field is limited.
- An electron beam with a large diameter d and a small angle α has a large focal spot with poor detail resolution (details are *blurred*). However, larger tube currents are possible, and a wide field of illumination can be achieved.

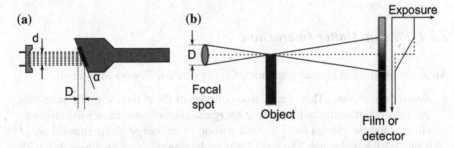

Fig. 2.7 Relevance of the focal spot size. **a** The anode angle α relates the focal spot (i.e., the diameter of the electron beam d) to the projected focal spot size D through $D = d/\tan\alpha$. The focal spot size limits the size of small details that can be discerned in the image. **b** With an idealized point source, the edge of any radio-opaque object leads to a steep transition of the exposure at the detector. When the focal spot is large, a transition zone appears, in which the sharp edge appears blurred on the detector

A small field of illumination is sometimes desirable, for example, for pencil-beam CT tubes. In many cases, the emphasis lies on a small focal spot, for example for dental X-rays. Several designs exist where the size of the focal spot can be adjusted, for example, by providing a cathode with two emitters (Wehnelt cylinders): a wide field illumination with a large focal spot can be used for scout images, and by switching to the fine-focus cathode, a sharper image of a smaller zone of interest can be taken.

2.2 X-Ray Attenuation

We will now examine the source of X-ray *contrast*: the absorption of X-rays in matter of different density and composition. We will first examine the mechanisms of absorption on the atomic scale, and then create a macroscopic model of absorption. High-energy X-ray photons lose some or all of their energy in a collision with atoms along their path. Often, the photon is deflected from its path, i.e., scattered. In addition, the high-energy photons can elevate shell electrons to a higher energy level, causing either *excitation* or *ionization*:

- *Ionization* occurs when an electron is ejected from the atom into the continuum. Since the electron is completely ejected, the atom is positively charged (an ion) until the electron vacancy is filled. Ionization requires energy (namely, the binding energy of the electron's shell), and filling the vacancy releases this energy again (photon).
- *Excitation* occurs when an electron—most often a k-shell electron—is elevated to an outer shell. Excitation requires energy, namely, the binding energy difference between the two shells. When the vacancy in the inner shell is filled, the energy is released again (photon). Since the electron remains in one of the atomic shells, the excited atom does not carry a charge (i.e, does not become an ion).

2.2.1 Photon-Matter Interaction

An X-ray photon can interact with matter in any of the following four ways:

1. *Rayleigh scattering*. This is an elastic collision of the photon with the atom that occurs predominantly at low photon energies. The collision causes the atoms to vibrate, and the photon loses a small portion of its energy. Scattering occurs at a small deflection angle. The rate of Rayleigh scattering rapidly diminishes with increasing X-ray energy, and Rayleigh scattering plays a negligible role in the diagnostic range from 50 keV upward.
2. *Compton scattering*. Compton scattering is the dominant interaction in a wide range of photon energies. In this type of interaction, the X-ray photon ejects an electron (ionization), and the X-ray photon loses the amount of energy needed to ionize the atom. The photon can be deflected by a large angle. Filling of

the shell vacancy generates additional X-ray photons that lead to the typical energy peaks of characteristic radiation. These photons are emitted in a random direction. Compton scattering is undesirable in the imaging process, because it causes ionization along the X-ray path, and because of the unpredictable large-angle deflection of the photon, which leads to background haze and therefore lowered contrast.

3. *Photoelectric effect.* The photoelectric effect is the most important interaction for image formation. In this type of interaction, the energy of the X-ray photon is *completely* expended for ejecting and accelerating an electron (ionization) as sketched in Fig. 2.8. The kinetic energy of the electron is $m_e v^2/2 = E_{\text{photon}} - E_{\text{shell}}$. The ejected electron can have a high kinetic energy and cause additional ionization (Auger electron) and X-ray production through deceleration. Filling of the ion vacancy generates characteristic radiation. The photoelectric effect diminishes rapidly with higher energies, but dominates in the diagnostic range from 50 to 70 keV. The X-ray absorption rate of the photoelectric effect depends directly on the density of the matter. Moreover, the photon is not scattered. Therefore, the photoelectric effect is the most important contributor to X-ray contrast.

4. *Pair production.* At very high energies (above 1.02 MeV) the photon can spontaneously transform into a positron–electron pair upon colliding with an atom. This process is a total energy-matter conversion, and the energy needed is the equivalent of two electron masses: $E_{\text{photon}} = 2m_e c^2$. Since the energy needed for pair production is far above the diagnostic energy range, pair production does not play a role in diagnostic imaging.

In Fig. 2.9, we can see the contribution of the individual scattering events to the total absorption. At low energies, the total attenuation coefficient is very high, and most of the low-energy radiation gets absorbed by the patient. At high energies, Compton scattering becomes dominant, leading to poor contrast and to haze. A good energy range for diagnostic imaging is in the range from 50 to 80 keV, and for some applications, such as CT, even up to 150 keV.

The ability of X-ray photons to create ions along their path—either through Compton scattering or through the photoelectric effect—is the cause of radiation damage and radiation burns. Molecules, particularly organic molecules, become more reactive when one or more of their atoms are ionized. These reactive species have a higher tendency to break apart or to react with nearby molecules. Cells can repair a certain amount of damage and therefore resist the influence of background radiation and low levels of radiation exposure. When radiation exposure exceeds the ability of the cell's repair mechanism, cells sustain irreparable damage and are replaced much like normal wound healing. If the damage exceeds the self-healing ability of the tissue, the tissue or organ may fail. Independently, ionization puts DNA elements at risk of breaking apart, which increases the probability of cancer development.

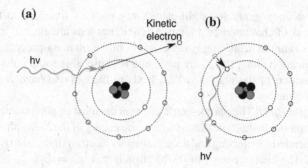

Fig. 2.8 Illustration of the photoelectric effect. **a** A high-energy X-ray photon expends all of its energy by elevating an inner-shell electron into the continuum. The difference between the photon energy and the binding energy becomes the kinetic energy of the electron. Note that this electron can have a high enough kinetic energy to cause braking radiation X-rays. Furthermore, the inner-shell vacancy is filled (**b**), which releases characteristic -rays

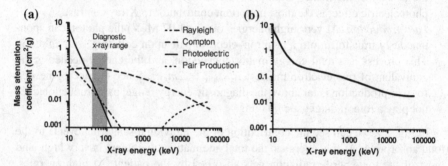

Fig. 2.9 **a** Contribution of the individual absorption events to the total X-ray absorption in a tissue-equivalent material, and **b** total absorption as the sum of the four contributing absorption events. Rayleigh scattering and photoelectric effect dominate at small energies. At extremely high energies far above the diagnostic energy range, pair production becomes dominant. Compton scattering is present over a wide range of energies. The ideal energy range for diagnostic purposes (highlighted in *gray*) exists where absorption is low enough for the majority of the photons to pass through the object, but where the photoelectric effect still contributes strongly to the total absorption. The energy-dependence of the total absorption can clearly be seen

2.2.2 Macroscopic Attenuation and Lambert-Beer's Law

We know that photons get absorbed when they pass through matter. On a macroscopic scale, the absorption manifests itself as the linear attenuation coefficient μ, usually given in cm^{-1} or the mass attenuation coefficient μ/ρ, given in cm^2/g. The two coefficients are related by the material density ρ. The absorption behavior is governed by Lambert-Beer's law, which can be derived by considering a very thin slice of absorbing material (thickness $\Delta x \to 0$). From N incident photons, Δn photons are absorbed. Δn is proportional to (a) the number of incident photons, (b) the thickness

Δx, and (c) the linear absorption coefficient μ:

$$\Delta n = N \cdot \mu \cdot \Delta x \tag{2.6}$$

If the slice is thick, N becomes a function of x, because each infinitesimal layer of material has its own incident number of photons, $N(x)$. In other words, the number of absorbed photons $dN(x)/dx$ is proportional to the number of incident photons at that location, $N(x)$, and the absorption coefficient μ. We obtain a first-order differential equation for N:

$$\frac{dN(x)}{dx} = -\mu \cdot N(x) \tag{2.7}$$

To solve the differential equation, we rearrange Eq. 2.7:

$$\frac{dN(x)}{N(x)} = -\mu \cdot dx \tag{2.8}$$

Now, Eq. 2.8 can be integrated from $x = 0$ to the length of the material l under the boundary condition that the incident number of photons at $x = 0$ is N_0:

$$\ln N(l) - \ln N_0 = -\mu l \tag{2.9}$$

Solving Eq. 2.9 for $N(l)$ yields Lambert-Beer's law [1]:

$$N(l) = N_0 \cdot e^{-\mu l} \tag{2.10}$$

The attenuation coefficient is always a function of the photon energy, that is, $\mu(E)$. The energy-dependency of the mass attenuation coefficient for three relevant tissues (muscle, adipose tissue, and bone) is shown in Fig. 2.10. In many cases, it is sufficient to approximate the tissue by some effective attenuation coefficient. However, for the quantitative determination of μ by absorption measurement, beam hardening effects need to be taken into account. *Beam hardening* refers to a shift of the peak energy toward higher energies due to stronger absorption of X-ray photons at lower energies. In the example of Fig. 2.11 with aluminum filters of 2, 5, and 10 mm thickness, the overall photon flux is reduced by 21, 39, and 59 %, respectively. However, the photon flux of photons above 80 keV is only reduced by 9, 21, and 38 %, respectively. The shift of the energy peak toward higher energies can clearly be seen.

Beam hardening effects are of minor importance in planar X-ray imaging, but can cause major artifacts in computed tomography. For this reason, beam hardening filters are used that absorb some of the low-energy radiation (see Fig. 2.6). Beam

---

[1] In optics, the base-10 logarithm is frequently used (e.g., the molar extinction coefficient ε of a fluid), whereas X-ray absorption coefficients μ use base-e logarithms. This is a common source of confusion.

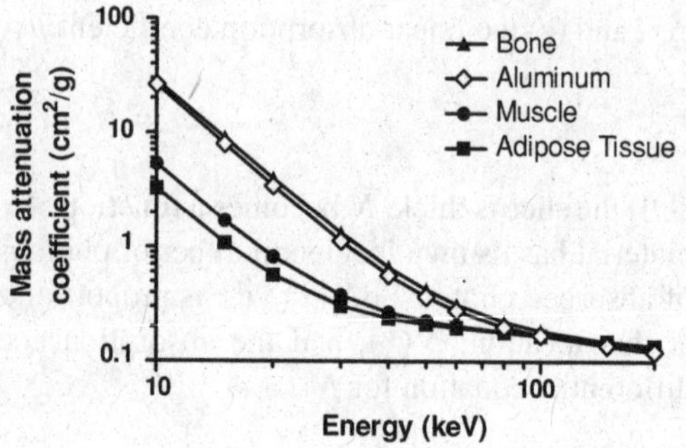

Fig. 2.10 Mass attenuation coefficient μ/ρ for muscle, adipose tissue, compact bone, and, for comparison purposes, aluminum [8]. X-ray imaging shows superior contrast between tissue and bone, but contrast between different types of tissue is relatively low

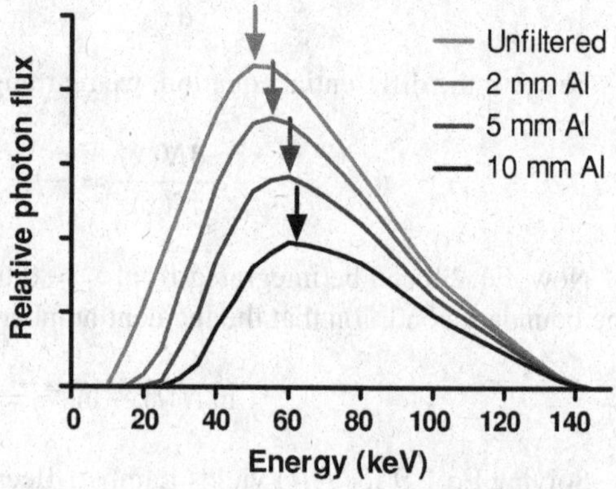

Fig. 2.11 Effect of beam hardening. Shown is the sample spectrum (characteristic X-rays omitted) of a tube at 140 keV (Fig. 2.5) and the filtered spectra after 2, 5, and 10 mm of aluminum. Arrows highlight the energy maximum.

Table 2.2 Linear attenuation coefficients (approximate) of some materials and biological tissues at 50 keV

Tissue/material	Linear attenuation coefficient (cm^{-1})
Air	0.00029
Water	0.214
Blood	0.241
Adipose tissue	0.193
Muscle	0.226
Brain	0.237
Compact bone	0.573

hardening filters are thin films of metal that preferentially absorb the lower-energy photons.

Table 2.2 lists the linear attenuation coefficients of some materials and biological tissues at 50 keV. The large difference between bone and other biological tissues is particularly prominent. The absorption coefficients of other tissues, e.g., blood, muscle tissue, or brain matter, are very similar, and X-ray imaging provides excellent bone-tissue contrast, but poor tissue-tissue contrast.

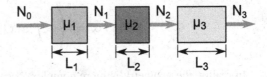

Fig. 2.12 X-ray absorption in three blocks of material of different size and different absorption coefficients

2.2.3 Lambert-Beer's Law in Inhomogeneous Materials

Equation 2.10 assumes a constant absorption coefficient along the path length l. What does Lambert-Beer's law look in inhomogeneous materials? Figure 2.12 shows an X-ray beam attenuated by three consecutive blocks of material with different length L_i and different absorption coefficient μ_i with $i \in [1, 2, 3]$. If we assume the incident number of photons to be N_0, we can compute the non-absorbed photons N_1 entering the second block:

$$N_1 = N_0 \cdot e^{-\mu_1 L_1} \qquad (2.11)$$

For the second and third block, we apply Eq. 2.10 in a similar manner:

$$N_2 = N_1 \cdot e^{-\mu_2 L_2}$$
$$N_3 = N_2 \cdot e^{-\mu_3 L_3} \qquad (2.12)$$

By combining the three attenuation equations into one, we can see that the exponential factors are multiplied, or the exponents added:

$$N_3 = N_0 \cdot e^{-\mu_1 L_1 - \mu_2 L_2 - \mu_3 L_3} \qquad (2.13)$$

It does not matter if the materials are separated by air gaps or contiguous—the overall absorption remains the same. For the continuous case of attenuation by an inhomogeneous material along a path s, we can write the general form of Lambert-Beer's law as

$$N = N_0 \cdot \exp\left(-\int_s \mu(\sigma)\, d\sigma\right) \qquad (2.14)$$

where $\mu(\sigma)$ is the attenuation coefficient at any point σ of the path s.

We need to understand X-ray imaging as *projection imaging*. The X-ray image is two-dimensional, and information in one spatial dimension is lost. Let us assume that the object (patient) can be described by the absorption $\mu(x, y, z)$ as a function of the three spatial dimensions x, y, and z. Let us further make the two simplifying assumptions that (1) the image plane is the x-y-plane, and (2) that the X-ray illumination is homogeneous and approximately parallel to the z-axis. In this case, the image $I(x, y)$ that represents the detector exposure relates to the object through

$$I(x, y) \propto \exp\left(-\int_z \mu(x, y, z)\, \mathrm{d}z\right) \tag{2.15}$$

In reality, the geometry is more complex, because the X-ray source emits rays in a cone-shaped configuration, and scattered photons create a background haze that reduces overall contrast. Moreover, the incident intensity is inhomogeneous within the cone. Lastly, the detectors, most notably film, often have a nonlinear characteristic. For these reasons, conventional X-ray imaging is widely qualitative, and it requires an experienced radiologist to identify the features of interest and make a diagnosis.

2.2.4 Dual-Energy X-Ray Absorptiometry

One area where projection X-ray imaging is used for highly accurate quantitative measurements is DEXA (dual-energy X-ray absorptiometry), usually used to measure bone density to diagnose osteopenia or osteoporosis. DEXA scanners are accurately calibrated, and the incident photon flux is either homogeneous or its spatial distribution known. DEXA operates under the assumption of a two-component model, i.e., that X-rays are attenuated along the path by either bone or soft tissue: When an X-ray beam travels through both bone and soft tissue, Eq. 2.10 can be extended to reflect X-ray attenuation by both tissues,

$$\frac{I}{I_0} = \exp\left(-\mu_B D_B - \mu_S D_S\right) \tag{2.16}$$

where μ_B is the X-ray absorption coefficient for compact bone and μ_S is the soft tissue absorption coefficient. It is assumed that the X-ray travels a total length D_B in compact bone and D_S in soft tissue, irrespective of the order of the tissues (such as, for example soft tissue—bone—soft tissue). Usually, D_S is not known and X-ray absorption by soft tissue causes a higher apparent bone density. The unknown absorbance can be eliminated by measuring the intensity at two different energies along the same path. The X-ray attenuation coefficients are energy-dependent (Fig. 2.10), and two different intensities I_H and I_L are obtained from high- and low-energy X-rays, respectively:

$$\begin{aligned}
I_H &= I_0 \exp\left(-\mu_{B,H} D_B - \mu_{S,H} D_S\right) \\
I_L &= I_0 \exp\left(-\mu_{B,L} D_B - \mu_{S,L} D_S\right)
\end{aligned} \tag{2.17}$$

When the incident intensity is known, for example, from an X-ray path through air, the intensities in Eq. 2.17 can be converted into total absorbance A_H and A_L, respectively:

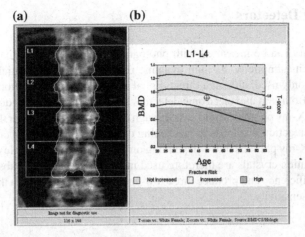

Fig. 2.13 DEXA scan of the lumbar spine. Lumbar vertebrae L1 through L4 haven been segmented by the software (**a**), and bone density is determined for the outlined regions. Since density is averaged for each vertebra, the low spatial resolution seen in the AP projection is acceptable for DEXA. The software also displays the average density in relation to a standard cohort (**b**). The three lines indicate mean bone density over age (*central line*) and the standard deviation of the cohort (*upper and lower lines*). Bone density for this patient was found to be slightly below the age-matched mean. *Image* © Springer Verlag, 2011, from Haidekker MA & Dougherty G, in: Medical Image Processing: Techniques and Applications, Springer 2011

$$A_H = \mu_{B,H} D_B + \mu_{S,H} D_S$$
$$A_L = \mu_{B,L} D_B + \mu_{S,L} D_S \tag{2.18}$$

The unknown quantity D_S can now be removed from the absorbance image A by computing a weighed difference of both absorbances,

$$A = \mu_{B,H} D_B + \mu_{S,H} D_S - w \left(\mu_{B,L} D_B + \mu_{S,L} D_S \right) \tag{2.19}$$

where w needs to be chosen as

$$w = \frac{\mu_{S,H}}{\mu_{S,L}} \tag{2.20}$$

Like conventional X-ray imaging, DEXA is a projection imaging method. Image analysis steps consist of the automated detection of the bone region and the computation of the averaged density. An example is shown in Fig. 2.13. DEXA is typically applied to the thoracic or lumbar spine, the femoral neck, or the calcaneus.

The accuracy of the DEXA method is limited for two main reasons. First, the absorption coefficients are approximations, because the X-ray beam is polychromatic. Second, and more importantly, the premise of the DEXA model is a two-tissue system (bone and soft tissue). However, soft tissue may be composed of muscle and adipose tissue, and the absorption values $\mu_{S,H}$ and $\mu_{S,L}$ show variability between individuals. In spite of these errors, DEXA typically shows a measurement accuracy of 10 % or better compared to ash mass [9].

2.3 X-Ray Detectors

X-ray intensity can be measured with photographic film or with digital detectors. Film, due to its simplicity, its high dynamic range, and its high spatial resolution is still commonly used, although direct digital acquisition becomes more and more widespread. Both film and electronic detectors have in common a low quantum efficiency. Most X-ray photons would pass through film, for example, rather than causing the chemical reaction that leads to blackening of the film. Scintillation crystals exist that convert X-ray photons into—depending on the application—visible or UV photons. Scintillation crystals are therefore used in conversion layers to absorb X-rays with a high efficiency and produce those low-energy photons for which the detectors show a high sensitivity.

2.3.1 Film-Based Imaging

Photosensitive film uses silver halide salts (usually $AgBr_2$) as the photosensitive compound. The silver halide is embedded in a thin, water-permeable layer (the *emulsion*). When exposed to visible or UV light, the salt breaks apart, and a *latent image* of elemental silver emerges. Two processing steps are necessary to create the final image. First, a developer (an aqueous, alkaline solution of several organic compounds) breaks apart more silver halide near the initial atoms of elemental silver. The elemental silver is responsible for the darkening (i.e., the high light absorption) of the exposed parts of the film. Second, a fixation solution (its central chemical is an inorganic thiosulfate) removes the remaining silver halide and prevents further darkening. Thorough rinsing is necessary to remove the development chemicals. The entire development process is usually performed by automated development machines—up to the level where the film is removed from the cassette by the machine, thus making a darkroom unnecessary (daylight processing).

For mechanical stability, the emulsion layer is supported by a transparent polymer film, the *carrier* (Fig. 2.14b). A thin, water-permeable layer mechanically protects the film on the emulsion side. Film has a very high dynamic range with light attenuation in the darkest regions of 1:1000 or higher. Some films have emulsion layers on both sides, effectively doubling their dynamic range.

Since film is sensitive towards light in the visible or UV range, the X-ray photons need to be converted. For this purpose, the film is placed in a cassette in close proximity to a conversion layer (Fig. 2.14a). Traditionally, tungsten compounds have been used, but newer materials based on rare earth metals show a higher efficiency (Table 2.3). Particularly, the high quantum yields of the rare earth materials allow a significant reduction of the patient dose. Calcium tungstate, for example, requires about three times the dose of terbium-doped gadolinium oxysulfide (Gd_2O_2S:Tb) to emit the same amount of light. The large difference in the absorption coefficient of

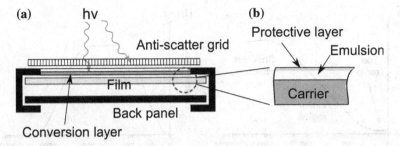

Fig. 2.14 Sketch of an X-ray film cassette (**a**) and photosensitive film (**b**). The cassette itself is a sturdy metal box with a removable panel to allow easy insertion of the photosensitive film. Most importantly, the cassette features a conversion layer, often on a glass support, that absorbs X-rays and emits visible or UV light. Film itself is composed of an emulsion layer that contains the photosensitive silver, applied onto a transparent polymer carrier

Table 2.3 X-ray absorption coefficient μ and quantum yield of some conversion layer materials

Conversion material	Absorption coefficient (mm^{-1})		Quantum yield (%)	Emission range
	at 40 keV	at 80 keV		
$CaWO_4$	4.00	3.15	4	Blue
$LaOBr : Tb$	13.1	1.86	13	Blue
$Gd_2O_2S : Tb$	4.62	3.29	19	Green

lanthanum oxybromide (LaOBr:Tb) between 40 and 80 keV can be explained with the presence of a k-edge slightly below 40 keV.

In a wide exposure range, film darkening follows a power-law relationship with the exposure. Since the human senses have logarithmic responses, film density is usually measured with the unitless logarithmic *optical density* OD, defined as

$$OD = -\log_{10} T = \log_{10} \frac{I_0}{I} \qquad (2.21)$$

where T is the transmission, that is, the fraction I of the incident illumination I_0 that is not absorbed by the film. We can approximately describe the film response to the X-ray exposure E as

$$OD = C \cdot \log E \qquad (2.22)$$

where C is the film's contrast. However, in both the high exposure range and the low exposure range, the film deviates from Eq. 2.22, and a typical film sensitivity curve is shown in Fig. 2.15. The film carrier and the emulsion are not perfectly clear. Films used in diagnostic radiology transmit about 78 % of the incident light (optical density of 0.11) when unexposed. Minor blackening of the film remains invisible against the natural fogginess of the carrier/emulsion system. Radiologists refer to the region of underexposure in the sensitivity curve as the *toe region* of the film. Conversely, there is an exposure level where all available silver halide has been converted to elemental

(a)
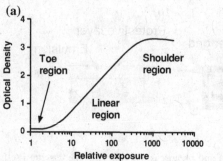

(b)
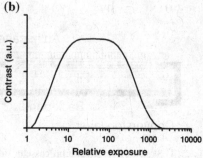

Fig. 2.15 Sensitivity curve **a** and contrast curve **b** of typical photographic film used in diagnostic radiology. In the toe region, very weak blackening cannot be seen in the natural fog and base absorption of the film. In the shoulder region, all available silver has been converted, and increased exposure does not lead to further blackening. A linear region between toe and shoulder exists. The film contrast is the first derivative of the sensitivity

silver, and further exposure does not increase blackening. This saturation region is called the *shoulder region*.

A wide variety of radiological films is available for any specific purpose. Films can be sensitized toward special conversion layers (for example, blue-or green-sensitive films), and films with different contrast are available. High-contrast films have a very narrow linear region and need to be accurately exposed. Films with lower contrast are good for general-purpose imaging and scout images. The region of optimal contrast (i.e., the region between toe and shoulder) is called the film's *latitude*. High-contrast films have a lower latitude than low-contrast films.

2.3.2 Fluoroscopes

The invention of the fluoroscope was the first major step in the direction of lower patient exposure and lower exposure times. Initially, a fluoroscope was a conversion layer (such as calcium tungstate) mounted in front of a light-shielded observation box. The radiologists used the fluoroscope to make the X-rays visible, but were exposed to the primary X-rays in the process. Development of image intensifiers remedied this situation.

Image intensifiers are based on a special photocathode composed of two conversion layers. The first layer is cesium iodide (CsI), which acts similar to the conversion layers in film cassettes, i.e., it emits visible light when hit by X-ray photons. However, cesium iodide has a very high quantum yield, producing about 3000 visible-light photons per X-ray photon (at 60 keV). Cesium iodide has the additional advantage of forming elongated crystals, and these crystals act as a form of natural anti-scatter grid. The second layer of the photocathode, often combinations of metals with antimony salts (Sb_2S_3), emits electrons when exposed to the visible light coming from

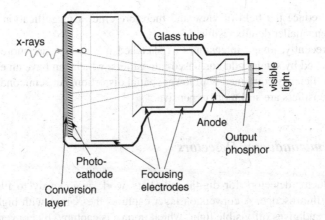

Fig. 2.16 Schematic representation of an electronic image intensifier. A conversion layer strongly absorbs X-ray photons and emits electrons. The electrons are accelerated and focused onto a luminescent output phosphor. The output window therefore displays a scaled-down and inverted image by visible-light emission. The image can be monitored in real-time with a TV camera and monitor

the CsI input layer. In vacuum, these electrons can be accelerated and directed onto a phosphor screen that is somewhat similar to the screen of a monochrome TV set, i.e., it emits visible light when hit by electrons. One example for an output phosphor material is ZnCdS:Ag, which emits green light when hit by kinetic electrons.

Figure 2.16 shows the principle of an image intensifier. A photocathode, preceded by a cesium iodide conversion layer, emits electrons when hit by an X-ray photon. These electrons are accelerated in an electrostatic field toward the anode. Typical acceleration voltages are 15–25 kV, similar to CRT TV tubes. A series of focusing electrodes ensures a highly defined path of the electrons such that each point on the input window is mapped to a point on the output window. For this reason, the output phosphor displays the image projected with X-rays onto the input conversion layer, but at a smaller scale and inverted. In addition, some geometric distortion occurs, and the output image is blurred to some extent. However these disadvantages are outweighed by the very high sensitivity of the image intensifier, which allows recording of the image in real-time, that is, at video frame rates. When the output phosphor is coupled to a video camera, the image can be monitored in a separate room.

Image intensifiers are fundamental for interventional radiology and image-guided surgery. In both situations, the patient is continuously monitored under low-dose X-ray illumination. X-ray source and image intensifier are mounted on a C-shaped arm, which allows flexible positioning with respect to the patient.

Moreover, fluoroscopy with image intensifiers allows special procedures, such as digital subtraction angiography, where an image before the injection of a contrast agent is subtracted from an image after contrast agent injection. The result is an image that displays only the contrast material-filled blood vessels and allows excellent visibility of the blood vessels. A change of the potentials on the focusing electrodes allows to reduce the field of view and therefore enter a magnification mode that makes even smaller details visible.

More recently, image intensifiers as sketched in Fig. 2.16 are more and more often replaced by semiconductor-based intensifiers, which can have an even higher quantum efficiency combined with lower spatial distortion, but semiconductor (flat-panel) intensifiers are still more expensive.

2.3.3 Semiconductor Detectors

Semiconductor detectors for digital radiology work analogously to film and the electronic fluoroscope: A conversion layer captures the X-rays with high quantum efficiency and gives off visible light, which in turn is captured by conventional visible light detectors, such as CCD (charge-coupled devices) and CMOS arrays. The basic light-sensitive element in a CCD detector is remotely related to a field-effect transistor with a metal-oxide gate. The main difference is the photosensitivity. Silicon p–n junctions are inherently sensitive to light, because a photon in the visible-light energy range can generate an electron-hole pair by elevating an electron from the valence band to the conduction band (photoelectric effect in semiconductors). In the special case of a CCD sensor, the channel under the gate is exposed to light. Electron-hole pairs are generated by light exposure, and the negative charges are held—and accumulated—under the influence of a positive gate potential. The process of exposure with charge accumulation and subsequent read-out is explained in Fig. 2.17. At the end of the channel, a charge amplifier and an A/D converter provide a digital signal that is proportional to the light exposure. CMOS sensors, which are more common than CCD sensors in low-end applications, follow a similar principle, but have added per-pixel processing circuitry at the expense of light sensitivity. For X-ray detection, the detector chip (CCD or CMOS) is sandwiched with a conversion layer. Unlike electronic image intensifiers, CCD-based semiconductor detectors are flat—even thinner than a film cassette. Unlike image intensifiers, spatial resolution can be very high. In fact, CCD chips with $10\,\mu$m pixel size or smaller are common. With a suitable microfocus X-ray tube, this gives rise to X-ray microscopy. On the other hand, CCD-based X-ray detectors generally do not achieve the sensitivity of image intensifiers, and a lower dose requires longer exposure times and simultaneously adds a strong noise component to the image.

Sensitivity can be increased by using the avalanche principle. In an avalanche diode, a low-doped intrinsic region is placed between the p- and n-Si zones (this configuration is called *pin-diode*). The intrinsic barrier allows the diode to be reverse-biased with high voltages of 100–200 V. Charges created in the junction zone by photoelectric absorption are accelerated by the high potential field and release additional charges along their path (thus, avalanche). Avalanche diodes are relatively large and require additional supporting circuitry. Although they can feature very high sensitivity, the spatial resolution is generally low, and avalanche diode imaging arrays are still a subject of active research.

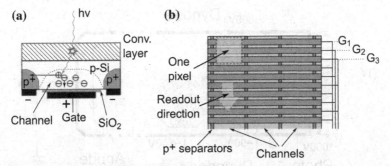

Fig. 2.17 Schematic of a charge-coupled device (CCD). The cross-section of a single element (pixel) is shown in **a**. In weakly doped p-Si, a n-channel is created by applying a positive voltage to a MOS gate. X-ray photons are captured by a conversion layer that is grown directly on top of the light-sensitive silicon wafer. The conversion layer emits visible light, which is captured by the silicon and creates an electron-hole pair. Electrons are captured and therefore accumulated in the positive gate field during the exposure, whereas holes (positive carriers) are shunted into the p^+ regions that separate the channels. **b** shows a view from the MOS gate side. Every third gate is connected. During exposure, only G_2 has a positive potential. During the read-out phase, a potential is applied to G_3, and the negative charges are attracted to the zone underneath G_3 and therefore move downward. Next, the positive potential is applied to G_1. Thus, the packet of accumulated charges is transported to the end of the CCD array, where it is read out. Each cycle G_2–G_3–G_1 allows to read out one row of pixels

2.3.4 Photomultiplier Tubes

The ultimate detection device when it comes to sensitivity is the photomultiplier tube (PMT). PMTs are high-voltage vacuum devices, somewhat like the electronic image intensifier. A PMT consists of several electron-multiplying *dynodes*. As shown in Fig. 2.18, an X-ray photon is absorbed by the conversion layer and converted into visible light. A visible-light photon that enters the PMT and hits the photocathode releases an electron—the photocathode is functionally similar to the photocathode of the image intensifier. This electron is accelerated in an electrostatic field toward the first dynode. In the process, the electron reaches a sufficiently high kinetic energy to release multiple electrons from the dynode it hits. These multiple electrons are now accelerated toward the second dynode, where *each* electron now releases multiple electrons, which in turn are accelerated towards the next dynodes. At the end of the dynode chain, the electrons are captured by the anode and create a measurable current spike *per photon* that hits the photocathode. Typical PMTs have 10 dynodes (multiplying stages), and each dynode releases between 3 and 5 electrons per received electron. Each photon can therefore cause a charge of 10 million electrons (1.5 pC) to be deposited on the anode, causing a pulse of several nanoamperes for a microsecond. The multiplication factor, that is, the average number of secondary electrons released per primary electron at any dynode is determined by the voltage difference between dynodes. Typically, a voltage divider creates the voltage gradient from the cathode to the anode, and by adjusting the cathode voltage, the overall sensitivity is controlled.

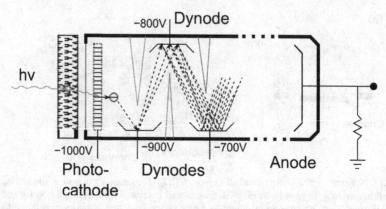

Fig. 2.18 Schematic of a photomultiplier tube (PMT). The light-sensitive part is very similar to the electronic image intensifier (Fig. 2.16) as an X-ray photon is converted into visible light, which in turn hits the photocathode, where an electron is released into the vacuum. This electron is accelerated toward the first dynode, where it gathers enough energy to release 3–5 secondary electrons from the dynode upon impact. These secondary electrons are accelerated toward the next dynode where the process is repeated. After several multiplying stages (i.e., dynodes), a measurable electron shower hits the anode

PMTs can be operated in a lower-sensitivity continuous mode, where the anode current is proportional to the photon flux at the photocathode, or in the higher-sensitivity *photon-counting* mode. In photon-counting mode, the current pulse that is caused by each photon at the anode is measured and counted. PMTs operating in photon-counting mode can easily be saturated (*pulse pile-up*), and in this mode, a PMT should not be operated above a flux 10^6 photons per second, because a strong deviation from linearity can be expected. High-sensitivity PMTs can be damaged by overexposure to light.

A very detailed insight into principles and practice of photomultiplier tubes can be gained from R.W. Engstrom's *Photomultiplier Handbook* [10], which is freely available on the web.

PMTs are relatively large, with optical windows between 5 and 50 mm. Therefore, any detector that relies on PMTs has a very low spatial resolution. However, this is a price worth paying when extremely low radiation doses are required, typically in nuclear medicine. In Chap. 4, we will see that PMTs are fundamental elements in the γ-sensitive *Anger camera* for SPECT (single-photon emission computed tomography) and for γ-detection in PET (positron emission tomography).

2.4 Factors that Determine X-Ray Image Quality

We have already encountered two factors that degrade X-ray image quality, *blur* and *haze*. Noise is an additional factor, notably in digital detection systems. In this section, the most important factors that cause image degradation are summarized.

- Geometric blur. Geometric blur occurs when details are imaged that are smaller than the X-ray beam diameter. The primary factor is the size of the focal spot, since the image is a convolution of the imaged object with the intensity distribution along the focal spot. Selectable focal spot sizes or special microfocus tubes help control geometric blur. Furthermore, keeping the object (patient) close to the detector also reduces geometric blur.
- Detector blur. Detector blur is caused by the conversion layer. The conversion layer typically emits visible light with equal probability in all directions. With a thick conversion layer, cross-talk between neighboring pixels is possible (whereas a thin conversion layer has a lower quantum efficiency). Keeping the conversion layer as close to the detector element as possible reduces detector blur.
- Motion blur. Motion blur occurs when the patient moves during exposure, for example by breathing. Generally, high detector sensitivity and high photon flux from the X-ray source allow shorter exposure times and reduce motion blur.
- Haze. Haze is primarily caused by scattered X-ray photons (Compton scattering). An anti-scatter grid reduces the amount of off-angle photons. For thin layers of tissue (e.g., the extremities), a lower X-ray energy can increase the influence of photoelectric absorption and Rayleigh scattering and thus decrease haze.
- Nonlinearity, over-and underexposure. Both over- and underexposure reduce image contrast. Overexposure applies equally to film and electronic detectors. Underexposure is more critical in electronic detectors due to the higher noise floor. Prior experience and—if necessary—a scout image can help selecting the correct exposure. The optical density of a film generally depends on the exposure in a nonlinear fashion. The use of calibration phantoms of known density allow per-image calibration.

Chapter 3
Computed Tomography

Abstract Computed tomography (CT), also known as computed axial tomography (CAT), is a volumetric imaging modality that is based on X-ray absorption. Unlike projection X-ray imaging (Chap. 2), CT allows the reconstruction of a two- or three-dimensional absorber map. CT vastly exceeds projection X-ray imaging in soft tissue contrast, but the spatial resolution of a clinical whole-body CT scanner is significantly lower than that of plain X-ray imaging. Nonetheless, CT can reveal small tumors, structural detail in trabecular bone or the alveolar tissue in the lungs. CT was introduced in 1971, and it is the first imaging modality where the computer is essential in the image reconstruction: a series of X-ray projections undergoes a transformation that yields the cross-sectional image. Since the introduction of the first CT scanners, major progress has been made in contrast, image quality, spatial resolution, and acquisition time. Modern clinical CT scanners are very fast and can produce a 2D cross-sectional image in less than a second. Spatial resolution can be as low as $100\,\mu$m in-plane, and specialized CT microscopes provide voxels of less than $10\,\mu$m. On the other hand, clinical CT scanners are expensive, ranging in the millions of dollars. This translates into a relatively high cost per CT scan, which prevents its more widespread adoption.

3.1 CT Image Formation Principles

In a projection image, such as a standard X-ray projection image, the exact location of an area of interest (e.g., lesion, tumor), cannot be determined. For example, the two objects (b) in Fig. 3.1a and b would generate similar X-ray projection images. For this reason, radiologists often take two perpendicular projections (e.g., lateral and AP = anterior—posterior), see Fig. 3.1c.

The goal of computed tomography (CT) is even more ambitious. The aim of CT is to obtain a spatially resolved map of absorption coefficients $\mu(x, y)$ in one slice of the patient's body. Such a map, if sampled at a finite resolution, is an image in our definition. The word tomography is a combination of the two Greek words for

M. A. Haidekker, *Medical Imaging Technology*, SpringerBriefs in Physics,
DOI: 10.1007/978-1-4614-7073-1_3, © The Author(s) 2013

(a)   **(b)**   **(c)**

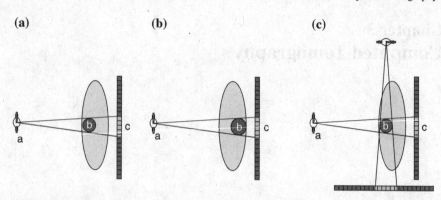

Fig. 3.1 Ambiguity in an X-ray projection. X-rays emitted by the source (*a*) are attenuated by an object (*b*, for example, a tumor) and cast a shadow (*c*) on the detector. The projection image is very similar in cases **a** and **b**, although the position of the tumor is very different. **c** Some of the ambiguity can be removed by taking two perpendicular X-ray images, such as an AP (anterior—posterior) projection and a lateral projection

Fig. 3.2 A block, composed of four unknown materials, can be probed with two perpendicular projections. Each attenuated beam intensity, I_a through I_d, provides an equation toward solving the unknown attenuation coefficients μ_1 through μ_4

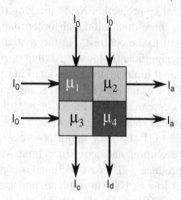

to cut and to graph: we obtain the view of a cross-sectional cut by using tomography without actually having to use a scalpel.

To understand the mathematical foundations that allow the reconstruction of a cross-sectional image from projections, let us look at a simplified example. A cubical object of 2 cm side length is composed of four equally-sized regions with different absorption coefficients μ_1 through μ_4 and the individual side length $d = 1$ cm (Fig. 3.2). We can obtain two different projections (lateral, AP) and determine the overall attenuation (attenuated beams I_a, I_b, I_c, I_d). Since each projection follows Lambert Beer's law, we obtain a linear equation system that we can solve for μ_1 through μ_4,[1]

---

[1] Unfortunately, even this simple equation system can only be solved when one of the absorption coefficients is known, because the four equations are not linearly independent

$$I_a = I_0 \cdot \exp\left(-\mu_1 d - \mu_2 d\right)$$
$$I_b = I_0 \cdot \exp\left(-\mu_3 d - \mu_4 d\right)$$
$$I_c = I_0 \cdot \exp\left(-\mu_1 d - \mu_3 d\right) \tag{3.1}$$
$$I_d = I_0 \cdot \exp\left(-\mu_2 d - \mu_4 d\right).$$

An arbitrary object, composed of n by n different materials requires n^2 independent equations, but two projections only provide $2n$ equations. The solution: We can take more projections at different angles. In fact, methods exist to solve a linear equation system of the type seen above. Since in the general case the number of equations and the number of unknowns do not match, numerical methods are employed to estimate the solution space of the linear equation system. These methods are referred to as ART (algebraic reconstruction techniques) [3] and are briefly covered in Chap. 4.

3.1.1 The Radon Transform and the Fourier Slice Theorem

The mathematical foundations for an efficient reconstruction algorithm were laid as early as 1917 by Austrian mathematician Radon [11, 12] who examined a form of integral transform that later became known as the *Radon transform*. It was almost 50 years later, however, that Radon's work was converted into a practical image reconstruction algorithm by Cormack [13, 14]. Cormack and Hounsfield, who counts as the actual inventor of the CT [7], shared the Nobel prize in 1979.

Let us define the straight line L as the set of all points $\vec{s} = (x, y)$ that obey the equation

$$x \cos\theta + y \sin\theta = \rho \tag{3.2}$$

where θ is the angle of the line with the x-axis and ρ is the distance of the line from the origin.[2] Furthermore, let us define the scanned object in the x, y-plane as $\mu(x, y) = \mu(\vec{s})$. We consider Lambert-Beer's law (Eq. 2.14) in its logarithmic form where a projection p is the X-ray absorption along its path, that is, $p = -\ln(N/N_0)$, where N_0 are the incident photons and N are the measured photons on the other side of the object. The X-ray beam follows the straight-line path L with the parameters ρ and θ. The projection $p(\rho, \theta)$ can therefore be expressed as

$$p(\rho, \theta) = \int_L \mu(s)\mathrm{d}s = \mathscr{R}(\rho, \theta)\{\mu\}. \tag{3.3}$$

This projection has become known as the Radon transform of μ for ρ and θ and is denoted $\mathscr{R}$. The geometry used in Eqs. 3.2 and 3.3 is illustrated in Fig. 3.3. The

---

[2] Here, the Radon transform is presented in 2D, because reconstruction most often takes place in a single plane (slice). The Radon transform can easily be extended into n-dimensional space

Fig. 3.3 Illustration of the
geometry in the Radon transform. An X-ray source emits
a pencil-beam ray, and some
of the X-ray photons get
absorbed as they pass through
the object with the absorption $\mu(x, y)$, and a detector
measures the intensity of the
X-rays passing through the
object. The pencil beam follows the straight-line path L
with angle θ and distance ρ
(Eq. 3.2). For any ρ and θ, the
detector can now determine
the Radon transform $p(\rho, \theta)$

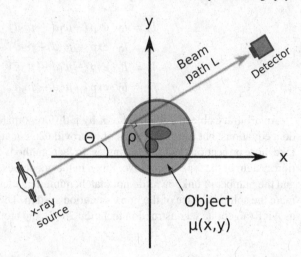

straight line L is the pencil-beam line from the X-ray source to the detector. X-rays
are absorbed as they pass through the object, which has the spatial distribution of
absorption coefficients $\mu(x, y)$. By translating the source-detector pair (i.e., changing
ρ) and by rotating it (i.e., changing θ), different Radon transform data $p(\rho, \theta)$ can
be collected.

In his seminal work, Radon has shown that the original function μ can be exactly
reconstructed from an infinite number of projections at every possible combina-
tion of ρ and θ [11]. In other words, an inverse transform $\mathscr{R}^{-1}$ exists for which
$\mu = \mathscr{R}^{-1}\{p\}$. The implication is that the projection data $p(\rho, \theta)$ can be used to
compute the unknown cross-section $\mu(x, y)$. The inverse transform is impractical,
however, because the infinite number of projections cannot be obtained. Numerical
methods need to be employed to account for the limited number of angular projections
and the placement of the beam at discrete steps. It is interesting to note that the Radon
transform can be elegantly expressed in the Fourier domain, where a direct relationship between the 2D Fourier transform of the sample, $M(u, v) = \mathscr{F}\{\mu(x, y)\}$ and
the 1D Fourier transform of the projections emerges. This relationship is known as
the *Fourier slice theorem* [3]:

> *The Fourier transform of a parallel projection of an image $\mu(x, y)$, taken at an angle θ, gives
> a 1D slice of the 2D Fourier transform $M(u, v)$ along a line through the origin $(u = 0,
> v = 0)$ and subtending an angle θ with the u-axis.*

The Fourier slice theorem can be derived in a very straightforward manner for any
projection parallel to one axis. With reference to Fig. 3.4, a projection $p(x)$ parallel
to the y axis can be defined as

$$p(x) = \int_{-\infty}^{+\infty} \mu(x, y)\, dy \qquad (3.4)$$

Fig. 3.4 Projecting the image $\mu(x, y)$ parallel to the y-axis yields one projection $p(x)$, where each point of $p(x)$ is the line integral along a *vertical line* (*arrow*) through $\mu(x, y)$

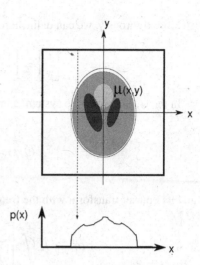

and its one-dimensional continuous Fourier transform $P(u) = \mathscr{F}\{p(x)\}$ is

$$P(u) = \int\limits_{-\infty}^{+\infty} \left[\int\limits_{-\infty}^{+\infty} \mu(x, y)\, dy \right] e^{-2\pi j u x}\, dx. \tag{3.5}$$

The two-dimensional Fourier transform $M(u, v)$ of the entire image $\mu(x, y)$ is defined as:

$$M(u, v) = \int\limits_{-\infty}^{+\infty} \int\limits_{-\infty}^{+\infty} \mu(x, y)\, e^{-2\pi j(ux+vy)}\, dx\, dy. \tag{3.6}$$

For $v = 0$ we obtain the 1D slice of the Fourier transform on the u-axis:

$$M(u, 0) = \int\limits_{-\infty}^{+\infty} \int\limits_{-\infty}^{+\infty} \mu(x, y)\, e^{-2\pi j u x}\, dx\, dy$$

$$= \int\limits_{-\infty}^{+\infty} \left[\int\limits_{-\infty}^{+\infty} \mu(x, y)\, dy \right] e^{-2\pi j u x}\, dx. \tag{3.7}$$

Since the exponential term is constant with respect to the integral over y, the equation can be rearranged and immediately becomes Eq. 3.5. It can now be argued that any image can be rotated by an arbitrary angle θ, and its Fourier transform is also rotated by θ. Therefore, any projection taken at an angle θ with the x-axis can be converted to a projection parallel to the y-axis by rotating both the image and its Fourier transform.

More rigorously, we can define a rotated (t, s) coordinate system:

$$\begin{bmatrix} t \\ s \end{bmatrix} = \begin{bmatrix} \cos\theta & \sin\theta \\ -\sin\theta & \cos\theta \end{bmatrix} \begin{bmatrix} x \\ y \end{bmatrix}. \tag{3.8}$$

In the (t, s) coordinate system, a projection along lines parallel to the s-axis is

$$p(\theta, t) = \int_{-\infty}^{+\infty} \mu(t, s)\, ds \tag{3.9}$$

and its Fourier transform with the frequency coordinate w

$$P(\theta, w) = \int_{-\infty}^{+\infty} \left[\int_{-\infty}^{+\infty} \mu(t, s)\, ds \right] e^{-2\pi j w t} dt. \tag{3.10}$$

By using Eq. 3.8, the Fourier-transform $P(\theta, w)$ of the projection can be transformed into the (x, y) coordinate system:

$$P(\theta, w) = \int_{-\infty}^{+\infty} \int_{-\infty}^{+\infty} \mu(x, y) e^{-2\pi j w(x\cos\theta + y\sin\theta)} dx\, dy. \tag{3.11}$$

The expression $x\cos\theta + y\sin\theta$ in the exponential term describes a straight line through the origin at an angle θ with the x-axis. Equation 3.11 can now be rewritten as the Fourier transform $M(u, v)$ of $\mu(x, y)$ with the constraints $u = w\cos\theta$ and $v = w\sin\theta$:

$$M(u, v) = \int_{-\infty}^{+\infty} \int_{-\infty}^{+\infty} \mu(x, y)\, e^{-2\pi j(xw\cos\theta + yw\sin\theta)} dx dy. \tag{3.12}$$

Equations 3.11 and 3.12 are the same, and we arrive at the equation for the Fourier slice theorem

$$P(\theta, w) = M(w\cos\theta, w\sin\theta). \tag{3.13}$$

3.1.2 Practical Image Reconstruction

Image formation and image reconstruction can most easily be explained with a CT scanner in *pencil-beam geometry*. Pencil-beam scanners, also known as first-generation scanners, have a highly collimated X-ray source and a detector positioned

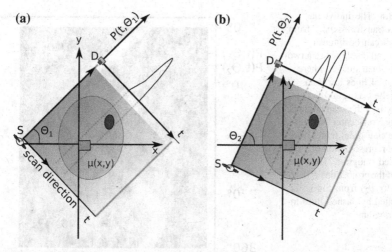

Fig. 3.5 Scanning process with a parallel-beam scan geometry. The scanned object is denoted $\mu(x, y)$. A pencil-beam source S emits an X-ray beam, and a detector D measures the X-ray intensity. **a** One projection $P(t, \theta_1)$ is acquired by moving the source-detector system along the t-axis perpendicular to the beam. The area covered by the scan is indicated by the gray gradient, and the projection profile $P(t, \theta_1)$ is drawn above the detector position. **b** For the next projection, the source-detector system is rotated by a small angle $\Delta\theta$ (now the angle is θ_2), and the projection scan is repeated. The new projection $P(t, \theta_2)$ looks different. Over 360° of rotation, a large number of projections is collected

on opposite sides of the sample. The source-detector system can be moved sideways (i.e., perpendicular to the beam direction) to cover the entire sample. Furthermore, the translation mechanism can be rotated with respect to the sample, but at discrete angular increments $\Delta\theta$ (Fig. 3.5). One scan, therefore, provides us with one intensity profile $I_{m,n}$ that is composed of discrete intensity samples, taken at regular intervals $t = m \cdot \Delta t$ along the t-axis, where t is perpendicular to the beam (cf. Eq. 3.9). The scans are taken at an angular position $\theta = n \cdot \Delta\theta$. The first step in the reconstruction process is to convert the intensity values into absorption values. The shoulders of $I_{m,n}$ left and right of the sample provide the unattenuated beam intensity I_0, and the discrete projection values are computed as

$$p_{m,n} = -\ln \frac{I_{m,n}}{I_0}. \tag{3.14}$$

The $p_{m,n}$ are the discrete representation of the projections $p(t, \theta)$. The measured absorption values can be arranged in a two-dimensional matrix, and the value of p shown as a proportional intensity (Fig. 3.6). This image representation of $p_{m,n}$ is referred to as *sinogram*, because inhomogeneities in the sample cause sinusoidal traces.

The Fourier slice theorem can now be used to devise an algorithm for image reconstruction from projection data. Such an algorithm to generate a reconstruction of the

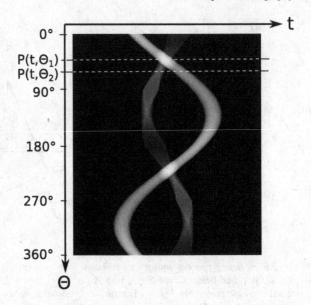

Fig. 3.6 The individual Radon transforms $p_{m,n}$ from Fig. 3.5 can be intensity-coded and displayed as a two-dimensional image, where horizontal lines are scans along the t-axis, and the vertical axis is the scan angle θ. Because off-center features create sinusoidal traces, the image representation of $p_{m,n}$ is called sinogram. The position of the projections $P(t, \theta_1)$ and $P(t, \theta_2)$ from Fig 3.5 is indicated by dashed lines in the sinogram

object (i.e, the attenuation matrix $\mu(x, y)$) would rely on the collection of many projections $p_{m,n}$. Their one-dimensional Fourier transforms $P_{w,n}$ are then entered into a two-dimensional frequency-domain matrix spaceholder that represents $M(u, v)$. Since it is practically impossible to fill all matrix elements of the frequency-domain matrix, missing elements need to be interpolated. Once the matrix is completely filled and interpolated, the cross-sectional image is obtained by two-dimensional inverse Fourier transform of the matrix.

The fundamental problem of this type of reconstruction is that a lot of information exists for the center of the 2D Fourier transform, while the information becomes sparse further away from the origin, requiring the interpolation of more missing values. Unfortunately, this information "further out", where most of the interpolated pixels lie, contains the high-frequency components, i.e., the image detail. This interpolation process is the single most important disadvantage of reconstruction in frequency space. Although special interpolation algorithms in frequency space exist, the most common approach today is the filtered backprojection (FBP) algorithm, which does not use the frequency domain at all. For FBP, the profile is backprojected ("smeared") into the original object plane along the projection angle. The backprojections of all profiles are added to form the reconstruction of the object.

Kak and Slaney [3] provided an elegant derivation of the backprojection motivated by the Fourier slice theorem. Since projection data are sampled in a polar coordinate system ($P_\theta(\omega)$ with the radial frequency ω and the sampling angle θ), the Fourier-domain spaceholder can be interpreted in polar coordinates as well, i.e., $M(\omega, \theta)$, and the inverse Fourier transform that yields the final image $\mu(x, y)$ from the two-dimensional Fourier data can be expressed in polar coordinates ω and θ:

$$\mu(x, y) = \int\limits_0^{2\pi} \int\limits_0^{\infty} M(\omega, \theta) e^{2\pi j \omega (x \cos\theta + y \sin\theta)} \omega \, d\omega \, d\theta \qquad (3.15)$$

where the following substitution has been made for the coordinates and the differential

$$u = \omega \cdot \cos\theta$$
$$v = \omega \cdot \sin\theta \qquad (3.16)$$
$$du \, dv = \omega \, d\omega \, d\theta.$$

A trick here is to split the outer integral into one integral from 0 to π and another from π to 2π, then shift the integration bounds of the second integral. Equation 3.15 can be rewritten as the sum of the two integrals,

$$\mu(x, y) = \int\limits_0^{\pi} \int\limits_0^{\infty} M(\omega, \theta) e^{2\pi j \omega (x \cos\theta + y \sin\theta)} \omega \, d\omega \, d\theta$$

$$+ \int\limits_0^{\pi} \int\limits_0^{\infty} M(\omega, \theta + \pi) e^{2\pi j \omega (x \cos(\theta + \pi) + y \sin(\theta + \pi))} \omega \, d\omega \, d\theta. \qquad (3.17)$$

We can make use of the symmetry of the 2D Fourier transform, $F(\omega, \theta + \pi) = F(-\omega, \theta)$, and a sign change of the integration bounds of the inner integral to join the two integrals and obtain

$$\mu(x, y) = \int\limits_0^{\pi} \left[\int\limits_{-\infty}^{\infty} M(\omega, \theta) e^{2\pi j \omega (x \cos\theta + y \sin\theta)} |\omega| \, d\omega \right] d\theta. \qquad (3.18)$$

In the Fourier slice theorem, the expression $M(\omega, \theta)$ is the one-dimensional Fourier transform of the projection $p(t, \theta)$, namely, $P(\omega, \theta)$ (cf. Eq. 3.10), which allows us to obtain the reconstructed image $\mu(x, y)$ directly from the projections as

$$\mu(x, y) = \int\limits_0^{\pi} \left[\int\limits_{-\infty}^{\infty} P(\omega, \theta) |\omega| \, e^{2\pi j \omega (x \cos\theta + y \sin\theta)} d\omega \right] d\theta. \qquad (3.19)$$

The expression inside the square brackets is a one-dimensional inverse Fourier transform of the projection, but with increasing frequencies weighted proportionally by $|\omega|$. We can define a *filtered projection* p_F,

$$p_F(t, \theta) = \int\limits_{-\infty}^{\infty} P(\omega, \theta) H(\omega) e^{2\pi j \omega t} d\omega \qquad (3.20)$$

where we made use of the line equation $t = x \cos\theta + y \sin\theta$ and generalized the function that emphasizes higher frequencies as $H(\omega) = |\omega|$. Thus, we now simply write the *filtered backprojection* as

$$\mu(x, y) = \int\limits_{0}^{\pi} p_F(x \cos\theta + y \sin\theta, \theta) d\theta. \qquad (3.21)$$

Why is Eq. 3.21 called *backprojection*? It clearly integrates all projections acquired over a semicircle of rotation. Let us look at the first projection where $\theta = 0$. The contribution for $\mu(x, y)$ is $p_F(x, \theta)$, and it contributes equally to all y. One could say, the projection is "smeared" along the y-axis. With the same rationale that we used to extend the Fourier slice theorem from a projection parallel to one axis to the general rotation, the projections at any angle θ are "smeared" over the place-holder $\mu(x, y)$ perpendicular to θ. This "smearing", or projection along one axis of a rotating coordinate system, is referred to as backprojecting.

The filter function $H(\omega) = |\omega|$ deserves some attention. If we use the unfiltered projection $p(t, \theta)$ in Eq. 3.21 (in other words, we set $H(\omega) = 1$), the reconstructed image would be degraded by considerable blur. More precisely, its point-spread function is approximately $1/r$ (see Sect. 1.3 for an introduction to the point-spread function). The consequence of this point-spread function is illustrated in Fig. 3.7 which clearly shows than even with a high number of projections the reconstruction appears blurred. By appropriately filtering the projection with the ideal filter $H(\omega) = |\omega|$, the $1/r$ point-spread function can be compensated for, and the reconstruction no longer appears blurred (Fig. 3.8).

The algorithm to use Eq. 3.21 for image reconstruction starts with the collection of projections $p_{m,n}$ as described at the beginning of this section. We can visualize the data as a sinogram, and each line of the sinogram is one projection. With the projection data $p_{m,n}$, we now perform the following steps:

Fig. 3.7 Image reconstruction by backprojection. **a** The projection of a cylindrical object is back-projected ("smeared") over an empty image placeholder. **b** Backprojection of two perpendicular projections. **c** With six projections, the original object begins to emerge. **d** Backprojection with 36 projections. The cylindrical object is clearly recognizable, but appears blurred

Fig. 3.8 Reconstruction process using filtered backprojection. The backprojections correspond to the second and fourth reconstruction in Fig. 3.7, but a filter has been applied. The blurred appearance has been corrected

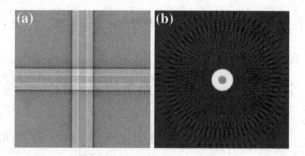

1. Prepare an empty image μ of m by m pixels and set all pixel values to zero. This is the placeholder for the reconstructed image.
2. For each line j of the sinogram $(0 \leq j < n)$,

   a. compute the filtered projection $p_F(m, n)$. One option is to compute the discrete Fourier transform, multiply each value with its frequency, and perform the inverse Fourier transform. The alternative is to perform a discrete convolution with a filter *kernel* as described below.
   b. for all pixels x, y of $\mu(x, y)$, compute $t = x \cos \theta_j + y \sin \theta_j$ where $\theta_j = j \cdot \Delta \theta$. Add $p_F(t, j)$ to $\mu(x, y)$. Note that t may be a non-integer value, and interpolation between the neighboring sample points is needed. Note also that due to the symmetry of the sine and cosine functions, x and y do not run from 0 to n, as conventionally done in image processing, but rather from $-n/2$ to $+n/2$.
3. Scale all pixel values by π/m.

Step 2b with the scaling factor in Step 3 represents the discrete approximation of the backprojection integral (Eq. 3.21), that is,

$$\mu(x, y) = \frac{\pi}{m} \sum_{j=0}^{m-1} p_F \left(x \cos \frac{\pi j}{m} + y \sin \frac{\pi j}{m}, j \right) \qquad (3.22)$$

where the scaling factor $\pi/m = \Delta \theta$ is the discrete equivalent of the differential $d\theta$ in Eq. 3.21 for a sinogram with m projections.

Special considerations can be taken into account when designing the filter for the filtered backprojection. The idealized filter that most accurately represents the scanned object has the frequency response $H(\omega) = |\omega|$ as shown above. Multiplying with the frequency response of a filter in the frequency domain corresponds to a convolution in the spatial domain. A discrete filter with the coefficients h_k can be designed that approximates the desired frequency response. In the context of CT reconstruction, these filter coefficients are referred to as the *reconstruction kernel*. The filter function in Eq. 3.23 has been proposed to approximate $H(\omega) = |\omega|$ [15]:

$$h(t) = \frac{\sin \pi t}{2\pi t} - \frac{1 - \cos \pi t}{2(\pi t)^2} \qquad (3.23)$$

which leads to the kernel values for the discrete convolution $h_0 = 1/4$, $h_k = -1/(\pi^2 k^2)$ for all odd k and $h_k = 0$ for all even $k \neq 0$. We can therefore formulate the filtering step as the discrete convolution of the projection values p with the filtering kernel as

$$p_F(t, j) = \sum_{k=-l}^{+l} p(t - k, j) \cdot h_k \qquad (3.24)$$

where l is the kernel support. Because the kernel coefficients drop off rapidly with increasing k, the convolution can be made relatively short, for example with $l = 4$, which leads to efficient time-domain filtering.

The filter function (Eq. 3.23) and its frequency response are shown in Fig. 3.9. It can be seen that the filter strongly amplifies the high frequencies (edge-enhancing filter). At the same time, high-frequency noise components are also amplified. For this reason, filter kernels have been proposed that show some attenuation at higher frequencies. A widely used kernel was proposed by Shepp and Logan [16] with a frequency response of $H(\omega) = |\omega| \cdot |(\sin \pi \omega)/\pi \omega)|$ for normalized frequencies $-0.5 \leq \omega \leq 0.5$. Its filter coefficients are $h_k = 2/(\pi^2(4k^2 - 1))$. Both the filter function $h(t)$ and the frequency response of the Shepp-Logan kernel are shown in Fig. 3.9. In comparison with the kernel by Ramachandran and Lakshminarayanan in Eq. 3.23, it amplifies high frequencies less and is therefore a softer kernel with slightly less image sharpness, but better signal-to-noise ratio (SNR). Taking this idea further, Rieder [17] proposed a series of functions resembling Gaussian functions that the filter kernel in Eq. 3.23 gets multiplied with to attenuate its high-frequency

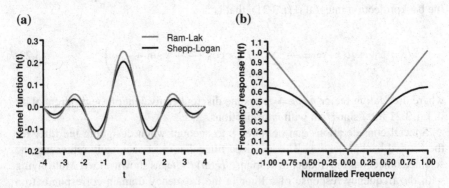

Fig. 3.9 Two widely used deconvolution kernels for filtered backprojection. **a** shows the filter function $h(t)$, and the filter coefficients h_k can be obtained from integer values of t. The resulting discrete convolution corresponds to a zero-mean finite-difference filter that approximates the first derivative. Their frequency response (**b**) shows that the Shepp-Logan kernel attenuates high frequencies compared to the kernel by Ramachandran and Lakshminarayanan and is therefore more robust against noise at the expense of some image sharpness

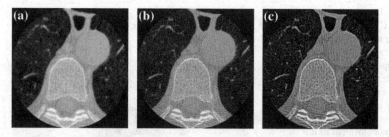

Fig. 3.10 CT reconstruction of a chest vertebra and its surrounding regions using different stock kernels. The vertebra, lung tissue (dark regions) as well as the circular aorta can clearly be identified. **a** Standard kernel, **b** bone kernel, **c** lung kernel. Each kernel amplifies specific structural aspects of the image

amplification, thus allowing to more flexibly balance reconstruction error against noise sensitivity.

Moreover, manufacturers often use proprietary modifications of the reconstruction kernel to enhance specific aspects of the reconstructed image. For example, General Electric scanners come with built-in kernels named after structural details that they are designed to enhance (Fig. 3.10). Specialized kernels like the ones named *bone* and *lung* emphasize higher spatial frequencies. On the other hand, these special kernels may also introduce artifacts, such as additional noise or pseudo-structure. A good example is the aorta as displayed in Fig. 3.10. Ideally, blood would show as a completely homogeneous region. However, both the *bone* and the *lung* kernel introduce some pseudo-texture that does not actually exist.

3.2 Engineering Aspects of CT Scanners

Pencil-beam scanners (also called first-generation scanners) generate a very thin beam of X-rays. A projection is generated by moving the source-detector system (source and detector on opposite sides of the patient) laterally alongside the patient (t-direction). For the next projection, the source-detector system is rotated around the patient (angle θ), and the next projection along t is obtained. This process is very slow due to the mechanical motion involved. Fan-beam scanners (also called second-generation scanners) reduce the time requirement by illuminating one entire slice of the patient (Fig. 3.11). The source emits a wedge-shaped beam which expands toward the patient and the detector. The detector is a one-dimensional array of sensors (line detector). The line represents the t-direction. Thus, a projection can be obtained without translational motion. Image reconstruction follows the same basic steps covered in the previous section, but the backprojection equation needs to be modified to account for the divergent beam geometry.

Fan-beam geometry is frequently employed for CT microscopes. Prerequisite is an X-ray tube with an extremely fine focal spot (in the order of few μm). By

Fig. 3.11 Sketch of a fan-beam CT scanner. The X-ray source and the detector array are arranged on a gantry on opposite sides of the sample. The source-detector system can rotate along the gantry (angle θ). The rotating coordinate t along the detector array is equivalent to that of a pencil-beam system, but a geometric correction for the divergent beam geometry is required in the reconstruction process

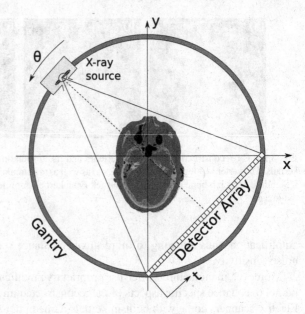

Fig. 3.12 Geometry of a fan-beam CT microscope (*top view*). By placing the sample close to the X-ray source, a magnification factor of b/a can be achieved. A microfocus X-ray tube is required to reduce geometric blur

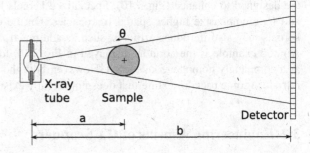

placing the sample close to the tube and relatively far away from the detector, a magnification of b/a can be achieved (Fig. 3.12). This arrangement causes some image degradation by scattered X-rays. Therefore, thorough collimation in the z direction is needed. Z collimation (slice thickness in z direction) also determines resolution in the z direction. Unlike patient CT scanners, CT microscopes often use a stationary source-detector system and rotate the sample inside the beam.

   A further extension of the fan-beam principle is to allow the X-rays to expand in a cone, much like the cone of a flashlight. With a two-dimensional detector array, more than one thin z slice can be illuminated and imaged in one projection. Acquisition time, especially for volumetric scans, can be significantly reduced. However, for the cone-beam geometry, image reconstruction becomes dramatically more complex. The most frequently used approaches are known as the Feldkamp backprojection [18], which is a 3D extension of the filtered backprojection, and the reconstruction proposed by Grangeat [19], which directly inverts the Radon transform.

A different scanning principle is used by the helical scanner (also known as *spiral CT*). A helical scanner moves the patient tray continuously while performing the source-detector revolution. In order to reconstruct one slice, projection points are obtained by interpolation in z direction so that the projection is perpendicular to the z axis. Two interpolation modes can be selected, *slim* and *wide* reconstruction. For *slim* reconstruction, projections that lie 180° apart are chosen, while 360° separation is used for *wide* reconstruction. Axial resolution is better in *slim* reconstructions at the expense of SNR. Helical scanners provide two major advantages: any slice position can be chosen for reconstruction without increasing patient radiation exposure, and motion artifacts are reduced over conventional CT.

3.3 Quantitative CT

Another important aspect of CT is the accurate reconstruction of the attenuation map $\mu(x, y)$ with respect to the X-ray absorption values μ. The X-ray spectrum is polychromatic, and the absorption is a function of the photon energy (see Fig. 2.10). Conventional reconstruction approaches use the attenuation along the beam path (Eq. 3.14) and assume a "representative" absorption μ for the Radon transform (Eq. 3.3), which will determine the image values returned in the reconstruction. The "representative" μ may vary widely between scanners, within the same scanner when different tube parameters are used, and even when the tube ages. A straightforward way to make the absorption coefficients more comparable between different scanners is to express absorption relative to water and air. Expressing absorption relative to water was originally proposed by Hounsfield [7], and the normalized attenuation values are named after him as *Hounsfield units* (HU). By definition, water is the reference point with 0 HU and air ($\mu = 0$) is defined as having -1000 HU. Equation 3.25 can therefore be used to convert specific reconstructed values $\mu(x, y)$ into Hounsfield units:

$$HU(x, y) = 1000 \cdot \frac{\mu(x, y) - \mu_{\text{water}}}{\mu_{\text{water}}}. \tag{3.25}$$

Typical HU values for different tissues are given in Table 3.1. Although contrast between different types of soft tissue is markedly better than in conventional X-ray

Table 3.1 Typical Hounsfield Units for different materials and tissues

Tissue	Relative attenuation (HU)
Fat	-200 to -50
Blood	40 to 60
Liver	20 to 50
Brain tissue	30 (gray matter) to 40 (white matter)
Bone	80 to 3000
Contrast agents	3000 and above

imaging, CT excels for imaging bone due to the high bone-tissue contrast and for the chest due to the high contrast between tissue and the air in the lungs. Contrast agents can be injected that increase the X-ray absorption of blood or of organs targeted by the contrast agents.

To improve accuracy, quantitative CT scans are often performed with several phantoms of known density in the field of view. Depending on the application, the phantoms approximate the density of known tissues or materials. A phantom could, for example, include a water-filled tube, another tube filled with an organic mixture that has a similar absorption coefficient as adipose tissue, and a resin with a known amount of calcium hydroxyapatite (bone mineral). In the cross-sectional image, average absorption coefficients of the phantom regions are measured. A linear regression with the known attenuation values of the phantom provides an accurate calibration curve.

3.4 Image Quality and Artifacts

Image quality is primarily determined by the detector size (spatial resolution) and the number of angular projections. Typical in-plane resolution of clinical CT scanners is 0.2–1 mm with slice thicknesses of 0.5–5 mm. Often, a higher slice thickness is chosen to reduce radiation dose and improve SNR at the expense of axial resolution.

The focal spot of the X-ray tube is another key determinant of image quality. A larger focal spot blurs the image (convolution of the ideal projection with the beam intensity profile). This is a major aspect in high-resolution CT and CT microscopes, where special tubes with 3–5 μm focal spot size are employed.

CT units, very much like X-ray detectors, employ collimators. The interaction of X-rays with tissue creates randomly scattered photons (e.g., Compton scattering)— randomly scattered photons create image noise and haze. Collimators in front of the detector act as anti-scatter grids and eliminate X-ray photons that deviate from a straight source-detector path. In addition, collimators in front of the X-ray tube can be used to reduce the beam size, to limit beam thickness (axial slice resolution), and to create a small apparent focal spot.

The detectors are a critical element of the unit. Detectors need to be centered with respect to the source; otherwise, the image will be blurred. In addition, detectors need to be calibrated so that the output signal intensity is identical for all detectors with the same incident X-ray intensity. Uncalibrated detectors create ring artifacts in the reconstruction. In extreme cases, a detector element may fail giving a constant output signal. Detectors and their associated signal amplifiers are also the primary source of noise. Electronic noise-suppression filters and software noise suppression reduce the amount of image noise. Noise can also be reduced by the operator by choosing a larger slice thickness or lower resolution, provided that the loss of detail is acceptable.

Since X-ray sources cannot provide monochromatic beams, an artifact related to beam hardening is common. Beam hardening occurs when the X-rays pass through strongly absorbing materials. Softer (lower-energy) X-rays are absorbed more readily, and the energy peak shifts toward higher energies as the X-ray beam passes through tissue. In projections with strong beam hardening, absorption values are therefore underestimated. Beam hardening can be reduced by prehardening the beam. A thin metal plate (frequently made of molybdenum or tungsten) is placed in front of the X-ray source. As X-rays pass through the plate, softer X-rays are absorbed, and the prehardened beam is less subject to beam hardening inside the object. In extreme cases, the reconstruction algorithm needs to be adapted. Multi-pass algorithms have been proposed [20, 21] where a first reconstruction is generated using standard reconstruction methods. Based on this reconstruction, approximate beam hardening can be computed through a ray-tracing approach, allowing to compensate for beam hardening in the projections. Reconstruction from the corrected projections reduces the influence of beam hardening.

Partial volume effects occur whenever a pixel represents more than one kind of tissue. This is particularly relevant when a tissue boundary lies within a CT slice. Partial volume effects blur the intensity distinction between adjacent tissues. Higher resolution or sometimes repositioning the patient may reduce partial volume effects.

Motion blur occurs when the patient moves (e.g., breathes) during the scan of one slice. Motion blur cannot be corrected, but the risk of motion blur can be reduced with shorter acquisition times and with the use of the helical scanning principle. Often patients are asked to hold their breath during the scan, but detailed imaging of the heart clearly continues to pose problems.

Chapter 4
Nuclear Imaging

Abstract Nuclear imaging is related to X-ray and CT imaging in that it uses radiation. However, unlike X-ray based imaging modalities, radioactive compounds are injected into the body as radiation sources. These radioactive compounds are typically linked to pharmacologically active substances ("radiopharmaceuticals") that accumulate at specific sites in the body, for example, in a tumor. With either projection techniques or volumetric computerized image reconstruction, the spatial distribution of the radiopharmaceutical can be determined. In this fashion, metabolic processes can be imaged and used for a diagnosis. Three-dimensional reconstructions are obtained in a fashion similar to CT, leading to a modality called single-photon emission computed tomography (SPECT). A parallel technology, positron emission tomography (PET), makes use of positron emitters that cause coincident pairs of gamma photons to be emitted. Its detection sensitivity and signal-to-noise ratio are better than SPECT. Both SPECT and PET have a significantly lower resolution than CT with voxel sizes not much smaller than 1 cm. Often, SPECT or PET images are superimposed with CT or MR images to provide a spatial reference. One limiting factor for the widespread use of nuclear imaging modalities is the cost. Furthermore, the radioactive labels are very short-lived with half-lives of mere hours, and most radiopharmaceuticals need to be produced on-site. This requires nuclear imaging centers to have some form of reactor for isotope generation.

4.1 Radiopharmaceuticals

Both computed tomography and magnetic resonance imaging provide predominantly structural information. One exception is functional MRI, where blood oxygenation levels, and therefore metabolic activity, can be determined. Functional information of a highly specific nature can be obtained by combining tomographic imaging with radioactively labeled pharmaceuticals (known as *radiopharmaceuticals* or radioactive tracers). Here, the radioactive compound acts as the radiation source (unlike the

M. A. Haidekker, *Medical Imaging Technology*, SpringerBriefs in Physics,
DOI: 10.1007/978-1-4614-7073-1_4, © The Author(s) 2013

Table 4.1 Some radiopharmaceuticals, their radioactive label and their predominant use

Radio- pharmaceutical	Radioactive label	Half life	Typical application
Fluorodeoxyglucose	^{18}F	2 h	Glucose metabolism, particularly in the brain, because it crosses the blood-brain barrier
Sodium iodide	^{123}I	13 h	Thyroid
Pentetreotide	^{111}In	2.8 days	Neuroendocrine tumors (somatostatin analog)
Strontium chloride	^{89}Sr	50 days	Bone tumors
Apcitide	^{99m}Tc	6 h	Acute thrombosis (binds to platelets)
Pentetate	^{99m}Tc	6 h	Kidney imaging (renal perfusion)
Krypton	^{81m}Kr	13 s	Lung ventilation imaging

external X-ray source in CT), and tomographic techniques are used to reconstruct the spatially dependent concentration of the radioactive compound.

Generally, two different types of radiolabels are used: Gamma emitters for single-photon emission computed tomography (SPECT) and positron emitters for positron emission tomography (PET). Examples for gamma emitters are the metastable isotopes of krypton (^{81m}Kr) and technetium (^{99m}Tc). Examples for positron emitters are ^{18}F, ^{15}O, ^{13}N, ^{11}C, and ^{82}Rb. In some cases, the radioactive substance can be used directly. The gas ^{81m}Kr can be inhaled for immediate lung ventilation studies. More often, a specific physiologically relevant compound is labeled with a radioactive atom. One example is fluorodeoxyglucose, which is metabolized at sites of high glucose demand, for example, in the brain or in some tumors. The spatial distribution of fluorodeoxyglucose accurately reflects the glucose uptake in the body. By substituting the stable ^{9}F for the positron-emitter ^{18}F, local concentrations of glucose can be imaged with PET. Some representative examples of radiopharmaceuticals are provided in Table 4.1. The examples in Table 4.1 consistently show a short half-life of the radioactive decay. Fast radioactive decay is desirable, because it reduces the total patient exposure. For example, ^{18}F has a half-life of approximately 2 h. Therefore, only 1 % of the initially administered ^{18}F is left after 12 h, while 99% has decayed into the stable ^{18}O. In fact, the fraction of ^{18}F remaining is even lower, because the radiopharmaceutical typically gets excreted through the urine.

4.2 Production of Short-Lived Radioactive Tracers

The typical short half-life makes transportation of the radiopharmaceuticals impractical: they need to be produced on-site and on demand. Radiolabels require a nuclear reaction to convert a stable atom (more precisely, the nucleus) into an unstable or metastable isotope. The isotope is then bound to the target drug by a chemical reaction. Some isotopes can be harvested from nuclear fission reactors.

The neutron-induced fission of ^{235}U creates a large number of isotopes with mass numbers ranging from 70 to 160. They can be separated by mass spectroscopy or by chemical reactions. Simultaneously, stable atoms can be bombarded with neutrons or protons to obtain isotopes. The three following examples illustrate commonly used pathways to generate isotopes:

- Production of ^{125}I: Stable ^{124}Xe is converted into the isotope ^{125}Xe by neutron activation in a fission reactor. Isotope ^{125}Xe decays with a half-life of 17 h into ^{125}I.
- Production of ^{18}F: The conversion of stable ^{18}O into ^{18}F is a typical example of a cyclotron reaction. A cyclotron is a particle accelerator in which charged particles, such as protons, can be accelerated to energies in the GeV range, enough to overcome Coulombic repulsion from other nuclei. Nuclei bombarded with protons usually transmute into positron emitters.
- Production of ^{99m}Tc: The popularity of ^{99m}Tc may be attributable to its relatively easy production pathway. The precursor, ^{99}Mo is produced by uranium fission in large quantities. ^{99}Mo has a relatively long half-life of almost 3 days and continuously produces ^{99m}Tc. Ammonium molybdenate can be adsorbed into an aluminum oxide column. Upon transmutation of ^{99}Mo into ^{99m}Tc, the resulting pertechnetate radical ($^{99m}Tc\,O_4^-$) reacts with the sodium in saline to form sodium pertechnetate, which can be easily eluted. The generator itself is a tabletop device that mainly contains a heavily shielded alumina column.

In a subsequent step, the radiopharmaceutical is created with conventional chemistry. For example, a two-step process creates fluorodeoxyglucose from mannose trifluoromethanesulfonate. Any "hot chemistry" reaction needs to be fast, because the radiolabel decays during the reaction. Large PET facilities are equipped with a cyclotron and an attached chemical lab adjacent to the PET scanner.

4.3 Detector Systems and the Anger Camera

Nuclear imaging requires the use of extremely sensitive detection devices, because higher detection sensitivity allows reducing the patient dose. Detectors are sensitive to high-energy gamma photons, since the relevant decay events produce gamma radiation. Most detectors are based on photomultiplier tubes (PMT), which are introduced in Sect. 2.3.4. For gamma detection, the PMT is combined with a conversion layer and additional collimators. Common materials for the conversion layer include thallium-activated sodium or cesium iodide (NaI(Tl), CsI(Tl)), cadmium tungstate (CdWO$_4$), and bismuth germanium oxide (BGO, Bi$_4$Ge$_3$O$_{12}$). All of these materials emit visible light, and the number of visible-light photons increases with the energy of the incident gamma photon. NaI(Tl) enjoys high popularity, because it shows a high sensitivity (about 2.5 times higher quantum yield than CdWO$_4$) and a decay time of about 230 ns (about 20 times faster than CdWO$_4$). Furthermore, NaI(Tl) forms elongated crystals that act as natural collimators, although additional collimation

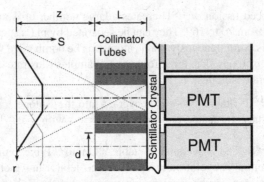

Fig. 4.1 Illustration of the spatial resolution of an array of collimated photomultiplier tubes (PMT). High-energy γ-photons are converted to visible light by a scintillation crystal (i.e., the conversion layer). Off-angle photons are attenuated by a collimator layer. The collimators can be thought of as cylindrical openings in a lead layer of length L and diameter d. The relative sensitivity s as a function of the radial position r from the center of the collimator tube is approximately trapezoidal and overlaps more and more with the sensitivity curve of the neighboring PMT (indicated in gray) with increasing distance z from the detector array

needs to be employed, because source collimation similar to the collimation of the X-ray tube is not possible. Detector collimation leads directly to the spatial resolution of a multidetector system. As sketched in Fig. 4.1, the sensitivity of a single PMT for a radiation source positioned off-axis from the collimator center decreases with the lateral distance r. The sensitivity function $s(r)$ widens with increasing distance z of the radiation source from the detector. Therefore, the sensitivity curve overlaps progressively more with that of the neighboring PMT as z increases. A full-width half-maximum criterion can be applied to obtain the half-maximum resolution Δr_H for cylindrical collimator holes [22],

$$\Delta r_H \approx 0.81 \cdot d \left(1 + \frac{z}{L} + \frac{h}{L} \right) \tag{4.1}$$

where h is the half-thickness of the scintillation layer. For high spatial resolution, a small d and large L is desirable, but the maximum sensitivity $s_{max} = d^2/16L^2$ follows the exact opposite trend in that it decreases with decreasing d and increasing L. It is not necessary to provide a single hole per PMT (as indicated in Fig. 4.1). Rather, collimators with multiple collimator channels per PMT are common and allow to better balance sensitivity and resolution requirements. Accordingly, a number of standard collimators exists for specific purposes, such as LEAP (low energy, all-purpose), LEHR (low energy, high resolution), or LEHS (low energy, high sensitivity) collimators.

Another important consideration is noise. In most imaging modalities, the noise component can be approximated as a Gaussian, zero-mean random variable. However, in nuclear imaging modalities, individual stochastic decay events (i.e., noise

itself) are recorded. In a simplified fashion, one could say that signal and noise are one
and the same. More precisely, the variance of the signal increases proportionally with
its average. This type of noise is called *Poisson noise*. In addition, the detector counts
both background radiation and gamma radiation that is scattered inside the patient
(cf. Sect. 2.2.1). Because of the low radiation levels involved, the PMT is operated
in its high-sensitivity photon-counting mode, and each pulse contains information
about the number of visible light photons that arrived simultaneously. Thresholding
of the PMT output pulse allows to suppress some of the lower-energy scattered photons. Integrating the PMT signal over some time improves the signal-to-noise ratio,
but the integration time is limited due to motion artifacts, the need to image larger
sections of the body, and the ongoing radioactive decay of the radiolabel.

If a large array of detectors, similar to the one sketched in Fig. 4.1, but extending
in two dimensions, is held above the patient's body, a projection image is obtained
in which the distribution of radiolabel concentration is resolved in two dimensions.
Such an array of detectors has been proposed by H.O. Anger [22, 23]. Modern
scintillation cameras, referred to as *Anger cameras*, are based on this design. Anger
proposed to arrange PMT detectors in a hexagonal pattern. Since neighboring PMT
cells have overlapping sensitivity functions (Fig. 4.1), the location $\vec{x} = (x, y)$ of a
radiation source is approximated by weighted summation of the signals S_i provided
by the individual detectors i,

$$\vec{x} = \frac{\sum_i \vec{x}_i S_i}{\sum_i S_i} \qquad (4.2)$$

where $\vec{x}_i$ is the spatial location of the ith PMT. The location of the event $\vec{x}_i$ (analogous
to the center of gravity) and the overall intensity $\sum_i S_i$ provide the image information
of intensity as a function of spatial location. If the camera is moved during recording,
the center coordinates of the moving camera are added to the center of gravity.
Weighted summation was performed by an analog network in the original designs.
However, modern designs use digital reconstruction techniques such as maximum
likelihood estimation [24, 25]. In scintigraphy, this information is often denoted as
(X, Y, Z) where X and Y refer to the centroid of the radiation source in a frame and
Z to its intensity.

In a practical application, a large-field Anger camera can be mounted on a gantry
that allows three-axis rotation and two-axis positioning (Fig. 4.2). By moving the
patient underneath the camera, a whole-body scintigram can be obtained. The planar
scintigram is the nuclear imaging analog of X-ray projection imaging.

4.4 Single Photon Emission Computed Tomography

Single Photon Emission Computed Tomography (SPECT) is the logical extension of
planar scintigraphy toward obtaining cross-sectional and three-dimensional images
of physiological activity, thus complementing the structural images obtained with

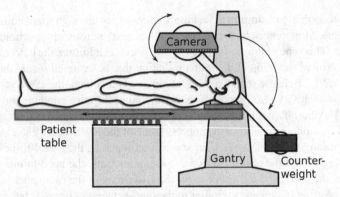

Fig. 4.2 Sketch of a whole-body scintillation imaging device. The Anger camera is mounted flexibly on a gantry to allow positioning of the camera close to the patient body. In addition, the camera can be placed at an angle with respect to the patient table to allow anterior-posterior, oblique, or lateral projections. By moving the patient table through the gantry, a whole-body projection image can be taken

Fig. 4.3 Sketch of a typical camera arrangement for SPECT. Multiple cameras are often used to capture as many radioactive events as possible, therefore to increase sensitivity and reduce scan time. Cameras are mounted on a gantry and can be rotated around the patient, thus capturing projections of the events along the camera's view direction (gray-shaded areas)

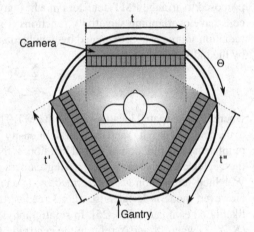

computed tomography or magnetic resonance imaging. At any one instant, an Anger camera takes a projection image of the radioactive decay along the camera's direction of view. If the camera can be rotated around the patient, as shown in Fig. 4.3, projections can be collected at different angles that lead to the same reconstruction principles used in CT. Compared to CT, however, SPECT has a much higher noise level and a lower spatial resolution.

The Anger camera allows the localization of a decay event by its centroid and event magnitude. The image can therefore be interpreted as the projection of the decay rate along a path p orthogonal to the camera's local (t, z) coordinate system. For reasons of uniformity, we denote as z the axis of the camera that is parallel to the direction of the patient table (pointing out of the picture plane in Fig. 4.3). For the recorded number of decay events at the camera $N(t, z)$, we can therefore use Eq. 4.3

as a starting point,

$$N(t, z, \theta) = \int\limits_{p} n(s, z) \, ds \qquad (4.3)$$

where $n(s, z)$ is the number of events occurring in the three-dimensional space defined by the path p and the axial position z. Equation 4.3 can be recognized as the Radon transform (see Sect. 3.1.1) with the path p as the set of all points (x, y) for which $x \cos \theta + y \sin \theta = t$ holds. In a SPECT device, the camera is therefore rotated around the patient's body to obtain multiple projections at different angles θ. Reconstruction takes place similar to computed tomography, and the camera's z direction provides multiple simultaneously acquired slices in the axial direction.

Equation 4.3 can be inverted with CT reconstruction methods, such as the filtered backprojection. However, the approximation suffers from two confounding factors. First, the number of recorded events depends on the strength of the source, $n(s, z)$ and on the distance ζ of the source from the camera. Furthermore, gamma radiation, like X-ray radiation, is attenuated inside the tissue. Equation 4.3 can be extended to include these terms:

$$N(t, \theta) = \int\limits_{p} \frac{n(s)}{\alpha \zeta^2} \cdot \exp\left[-\int\limits_{p'} \mu(s') \, ds' \right] \, ds \qquad (4.4)$$

Equation 4.4 has been simplified to a single axial slice (constant z). The constant α reflects the relative sensitivity of the camera, and the path p' is the part of p between the radiation source and the camera. No closed-term solution exists for the inverse problem of Eq. 4.4. The approximation in Eq. 4.3 is often sufficient, particularly when local concentrations of the radiopharmaceutical need to be detected, but the absolute activity level is of no critical concern. The filtered backprojection can be implemented in a robust fashion, but the filter function $H(\omega)$ would be chosen to attenuate high frequencies even more strongly than the filter functions discussed in Sect. 3.1.2.

Even more advantageous are iterative reconstruction methods, because those methods can be optimized to consider the Poisson noise. Iterative methods are based on a matrix formulation of the projection operation. Let us assume a two-dimensional slice with discretely distributed radiation sources that emit n_i photons. We further assume the slice to be composed of N by N pixels that are consecutively numbered, so that the index i runs from 0 to $N^2 - 1$. We can therefore take the n_i as elements in the emission vector $\vec{N} = \{n_0, n_1, \ldots, n_{N^2-1}\}$. We take M projections p_j, and the p_j are the components of the projection vector $\vec{P}$. Each projection takes place along a straight line that follows the line equation $x \cos \theta + y \sin \theta = t$ with the lateral position of the detector, t, and the angle of projection, θ. Projections are consecutively numbered from 0 to $M - 1$, and M is the product of the number of detectors (i.e., different possible t) and the number of angular positions. We can now interpret the event count in the Anger camera as the sum of all n_i along a straight line along

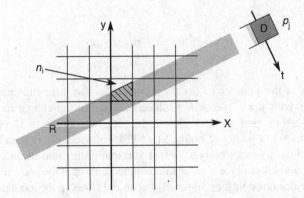

Fig. 4.4 Illustration of the system matrix w_{ij} in arithmetic reconstruction. A detector D records radiation from all locations along one ray R (diagonal gray region). The ray has the half-maximum width of the detector. The detector can be moved laterally (along the t-direction) and placed at different angles with respect to the x-axis to provide multiple projections p. Generally, the ray does not cover the entire area of one pixel. In this example, the contribution n_i of the radiation source in pixel i to the projection p_j is the hatched area. The fraction of the hatched area of the pixel area is w_{ij}

which the camera "looks". We can therefore write the projection as a *line sum*,

$$p_j = \sum_i w_{ij} \cdot n_i \qquad (4.5)$$

where w_{ij} is called the *system matrix*, which reflects the contribution of the ith pixel to the jth projection. In the simplest case, we can set $w_{ij} = 1$ for all i that are touched by the straight line j and $w_{ij} = 0$ otherwise. In practice, the relative contribution of a pixel of finite area to a ray of finite thickness is considered as sketched in Fig. 4.4. The individual contribution factors w_{ij} can be arranged in a matrix $\mathbf{W}$ with N^2 by M elements. The projection operation can now be written as the seemingly simple matrix operation

$$\vec{P} = \mathbf{W} \cdot \vec{N} \qquad (4.6)$$

At first glance it might appear as if the inversion of the sparse matrix $\mathbf{W}$ can provide the solution (i.e., the unknown pixel values n_i) through $\vec{N} = \mathbf{W}^{-1} \cdot \vec{P}$, but algebraic inversion is generally not possible. An iterative approach has first been proposed by Kaczmarz [26] and is known as algebraic reconstruction technique (ART).

For iterative reconstruction, the result vector $\vec{N}$ is initially filled with some initial value. Often $n_i = 0$ is used, but other choices can accelerate convergence. At any stage of the reconstruction, projections $\vec{P}'$ can be computed from the present solution vector $\vec{N}$. Generally, there will be an error between the measured projection and the projection of the current estimate, $\vec{E} = \vec{P} - \vec{P}'$. This error is backprojected over the solution vector. By using the system matrix w_{ij}, we can compute the individual elements of the error vector as

$$e_j = p_j - \vec{W}_j \cdot \vec{N} = p_j - \sum_i w_{ij} \cdot n_i \qquad (4.7)$$

where $\vec{W}_j$ indicates the jth row vector of the system matrix $\mathbf{W}$. The step to advance the solution from iteration k to $k+1$ (by using the error vector from the kth iteration) is to compute for each pixel n_i [1]

$$n_i^{[k+1]} = n_i^{[k]} + \frac{\sum\limits_j \dfrac{w_{ij} e_j}{\sum\limits_i w_{ij}}}{\sum\limits_j w_{ij}} \qquad (4.8)$$

Kaczmarz showed that the iteration converges. The iterative reconstruction process can be aborted either after a fixed number of iterations or after the error magnitude falls below a certain threshold.

The general reconstruction equation (Eq. 4.8) can be modified to reflect probabilities. Specifically, the Poisson nature of SPECT data leads to the question which image $\vec{N}$ has the highest *probability* to generate a set of projections $\vec{P}$ when the probability that a recorded count n_i was caused by a true activity n' has a Poisson distribution. The probability maximization leads to the *maximum likelihood expectation maximization* (MLEM) algorithm [27], in which the correction step is multiplicative rather than additive:

$$n_i^{[k+1]} = \frac{n_i^{[k]}}{\sum\limits_i w_{ij}} \cdot \sum_j w_{ij} \frac{p_j}{\sum\limits_l w_{lj} n_l^{[k]}} \qquad (4.9)$$

The denominator of the second fraction represents the projection of the reconstructed image in the present iteration: let us denote this projection of the reconstructed data as p'_j. The second fraction is the multiplicative error p_j / p'_j, which leads to the correction of the ith pixel n_i. The downside of the MLEM method is its tendency to produce noisy reconstructed images as the iterations progress. A variant of the MLEM algorithm is the *maximum a posteriori* (MAP) method, which differs from MLEM primarily by the introduction of an energy function E that penalizes a high noise component [28]:

$$n_i^{[k+1]} = \frac{n_i^{[k]}}{\sum\limits_i w_{ij} + \beta \dfrac{\partial}{\partial n_i} E(n_i)} \cdot \sum_j w_{ij} \frac{p_j}{\sum\limits_l w_{lj} n_l^{[k]}} \qquad (4.10)$$

---

[1] The step described here is actually referred to as SART for *simultaneous arithmetic reconstruction technique*, in which the complete error vector, rather than an individual error value, is backprojected.

The adjustable parameter β determines the strength of the regularization term, and it can be seen that for $\beta = 0$ Eq. 4.10 becomes Eq. 4.9. There are a number of formulations for the energy term, including the squared deviation of a pixel n_i from the average of its neighborhood or the squared deviation from the neighborhood's median value. To provide one example, we could define

$$\frac{\partial}{\partial n_i} E(n_i) = \sum_{l \in N3x3} (n_i - n_l)^2 \qquad (4.11)$$

where $l \in N3x3$ indicates the 3x3 neighborhood of pixel i in the conventional two-dimensional fashion.

Arithmetic reconstruction has only recently gained a foothold in SPECT reconstruction due to its computational expense. The filtered backprojection, for example, provides a reconstruction several orders of magnitude faster than ART or MLEM and remained the main-stream reconstruction method for a long time in spite of the superior image quality of iterative methods. However, the development of ordered-subset reconstruction and the availability of faster hardware (including accelerated processing on graphics hardware) are increasing the attractiveness of iterative reconstruction methods.

4.5 Positron Emission Tomography

Positron emission tomography (PET) differs from SPECT in one fundamental way: PET radionuclides are positron (β^+) emitters. After the decay event, the positron travels a very short distance before colliding with an electron. The ensuing matter-antimatter annihilation event creates two γ-photons with an energy of $E = m_e c^2 = 511$ keV each. The photons are traveling in exactly opposite directions. PET scanners use complete detector rings as sketched in Fig. 4.5. Individual detectors are PMTs with a conversion layer and a collimator, much like the individual elements of an Anger camera. As conversion materials, bismuth germanium oxide (BGO) is very common, but rare-earth materials, such as cerium-doped gadolinium oxyorthosilicate (GDO:Ce) and lutetium oxyorthosilicate (LSO:Ce) are gaining popularity due to their increased sensitivity and reduced luminescent decay time.

Unlike SPECT, the almost-simultaneous detection of two γ-photons can be expected, and single events can be rejected as noise. This method is referred to as *coincidence detection*. Furthermore, the line that connects the two PMTs that recorded the event serves as the direction of projection. As a result, PET has a dramatically better signal-to-noise ratio than SPECT. Coincidence detection requires ultrafast electronics. Any detector that records an event opens a gate for very short time window of length τ. If a second detector records an event during this window, a coincidence event is recorded; otherwise, the initial event is rejected. Noise is recorded when two independent events (this includes γ decay) occur within the

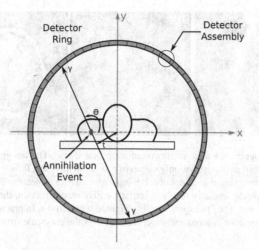

Fig. 4.5 Schematic representation of a PET detector ring. Multiple detectors are arrayed along the ring. Each detector assembly consists of a PMT, a scintillator crystal, and a collimator. A decay event inside the patient's body creates a pair of γ-photons traveling in opposite directions that are recorded by the detector ring. The event is known to have taken place somewhere along the line of angle θ with the x-axis and the distance t from the origin

detection window, or when a photon is scattered and changes its direction. Many PET scanners use τ in the range from 5 to 15 ns, during which light travels between 1.5 and 4.5 m. In theory, shorter τ could improve the coincidence detection with better noise rejection, but limits are imposed by the response time of the PMT and the decay time of the scintillator crystal. Coincidence detection also acts as a form of electronic collimation, and the collimators in PET scanners can be designed to be much smaller and weaker than those of SPECT scanners. As a result, the sensitivity of a PET scanner is several orders of magnitude higher than that of a SPECT scanner.

Image formation and reconstruction takes place by recording coincidence events over a certain period of time (integration). Each detector pair provides t and θ of the connecting line (Fig. 4.5), and the counts are collected as $n(t, \theta)$ for each detector ring individually. This is the sinogram of the PET scan (Fig. 4.6a). For image reconstruction, the same methods are used as for SPECT: filtered backprojection or, preferably, MLEM/MAP-based iterative reconstruction (Fig. 4.6b). Here, another difference between SPECT and PET emerges as a consequence of the coincidence detection. in PET, local attenuation does not influence the signal, only the total attenuation along a coincidence line. Attenuation correction is therefore more straightforward by either using an existing structural image (such as a CT image), or by simply assuming homogeneous absorption inside the body.

The reconstructed activity map is normally thresholded to eliminate background noise and then false-colored. The false-colored activity map is finally superimposed over a structural image from a CT or MRI scan (Fig. 4.6c).

Fig. 4.6 a PET sinogram of a slice with two radiation "hot spots". The sinogram can either allow angles θ from 0° to 360° with positive-only values of t, or, as in this case, limit θ from 0° to 180° and reflect larger angles with negative values of t. **b** Simple filtered backprojection-based reconstruction of the sinogram, showing spatially resolved activity. **c** PET reconstruction, thresholded, superimposed over a MR image. The "hot spots" are highlighted with arrows. In practice, the thresholded activity map would be false-colored and superimposed over the grayscale structural image

4.6 Multi-Modality Imaging

Both SPECT and PET provide functional, but not structural, information. It is often desirable to know the location of radiopharmaceutical accumulation as exactly as possible to relate the activity to a specific organ or site. For this reason, a concurrent structural image is often acquired with CT or MRI. The reconstructed activity map can then be superimposed over the structural image. Specialized dual-modality scanners exist, for example, combined SPECT-CT scanners, but they are not widely used because of their extremely high costs. Combining SPECT or PET with magnetic resonance is even more complex because of the high magnetic fields involved, and combined PET/MRI scanners are currently in the research stage [29].

More commonly, a structural CT or MRI image is taken, followed by a separate SPECT or PET scan. The separate acquisition of images immediately leads to the question of image fusion or *image registration*. First, SPECT and CT reconstructions typically have much lower resolution than MRI or CT images. Whereas a standard CT image can provide in-plane pixel sizes of 0.1–0.5 mm, PET and SPECT pixels are more in the range of 1 cm. Pixel and voxel sizes can be brought to match by interpolation. However, the position of the patient inside the scanner is highly variable, and some form of transformation is needed to bring both images to congruence [30]. In the simplest form, rigid-body transforms (rotation and translation), guided by external markers, can achieve congruence. The rigid-body model is usually sufficient for the head and long bones. Image registration of soft tissue, on the other hand, may involve more advanced methods of nonlinear transforms, such as using elastic models to account for soft-tissue deformation.

Chapter 5
Magnetic Resonance Imaging

Abstract Magnetic resonance imaging (MRI) is a volumetric imaging modality that parallels, to a certain extent, computed tomography. However, the underlying physical principles are fundamentally different from CT. Where CT uses high-energy photons and the interaction of photons with electrons of the atomic shell for contrast generation, MRI is based on the orientation of protons inside a strong magnetic field. This orientation can be manipulated with resonant radiofrequency waves, and the return of the protons to their equilibrium state can be measured. The relaxation time constants are highly tissue-dependent, and MRI features superior soft tissue contrast, by far exceeding that of CT. On the other hand, MRI requires dramatically more time for image acquisition than CT, unless special high-speed protocols are used (which often suffer from poor image quality). In addition, modern MRI scanners require a superconductive magnet with liquid helium cooling infrastructure, extremely sensitive radiofrequency amplifiers, and a complete room shielded against electromagnetic interference. For this reason, MRI equipment is extremely expensive with costs of several million dollars for the scanner hardware and with accordingly high recurring costs for maintenance. However, MRI scanners provide images with a very high diagnostic value, and MRI can be used to monitor some physiological processes (e.g., water diffusion, blood oxygenation) and therefore partly overlaps with nuclear imaging modalities. Since MRI is a radiation-free modality, it is often used in clinical studies with volunteers.

5.1 Proton Spins in an External Magnetic Field

Protons and electrons are charged particles, and charged particles in motion create a magnetic field. Electrons moving in a straight wire, for example, create a magnetic field H that drops off linearly with the distance r from the wire,

$$H = \frac{I}{2\pi r} \qquad (5.1)$$

M. A. Haidekker, *Medical Imaging Technology*, SpringerBriefs in Physics,
DOI: 10.1007/978-1-4614-7073-1_5, © The Author(s) 2013

where I is the current, that is, the number of unit charges per second. The magnetic field H has units of Amperes per meter (A/m). Better known is the *magnetic induction* B, which is related to the magnetic field through

$$B = \mu \cdot \mu_0 \cdot H \tag{5.2}$$

where μ_0 is the magnetic permeability in vacuum ($4\pi \times 10^{-7}$ Vs/A) and μ is the relative permeability of the material. The magnetic induction has units of T (Tesla = Vs/m). Vacuum and air have $\mu = 1$. Ferromagnetic materials have large relative permeabilities (examples: Steel, $\mu = 100$, ferrite, i.e., nickel-manganese-zinc alloys, $\mu > 600$, Mu-metal, i.e., alloys of nickel and iron with small amounts of copper and molybdenum, $\mu > 20000$). Ferromagnetic materials can exhibit *self-magnetism* when the magnetic domains are oriented in the same direction. Paramagnetic materials have a relative magnetic permeability greater than one, but much lower than ferromagnetic materials. Paramagnetic materials are attracted by magnetic fields, but do not exhibit self-magnetism, because thermal influences randomize the orientation of the magnetic domains in the absence of an external magnetic field. Paramagnetic materials of importance for MRI are oxygen and manganese, and the contrast agent gadolinium. Lastly, diamagnetic materials have a *negative* permeability ($\mu < 0$) and are repulsed by a magnetic field. Since carbon and hydrogen are diamagnetic, most organic compounds also have diamagnetic properties.

In the nucleus, protons *spin*[1] and thus create a magnetic moment $\vec{M} = \gamma \hbar \vec{I}$, where γ is the gyromagnetic ratio, $\hbar$ is the reduced Planck constant, and $\vec{I}$ is the proton's spin number with $I = 1/2$ for a proton. The gyromagnetic ratio relates the angular momentum to the magnetic moment and is a material constant. We use γ in the context of angular rotation and $\gamma/2\pi$ in the context of linear frequency. In MRI, hydrogen is the most relevant atom due to its abundance in organic tissue: One typical MRI voxel contains in the order of 10^{21} hydrogen atoms. The hydrogen nucleus consists of only one proton. The magnetic moment of a proton is very weak (Table 5.1), and the random orientation of the spins causes the magnetic moments in a small tissue volume to cancel out.

In an external magnetic field B_0, the spins experience a torque that aligns them with the direction of the external field as illustrated in Fig. 5.1a: A current loop (such

Table 5.1 Some properties of neutrons and protons

Characteristic	Neutron	Proton
Mass m	1.674×10^{-27} kg	1.672×10^{-27} kg
Charge q	0	1.602×10^{-19} C
Magnetic moment M	-9.66×10^{-27} J/T	1.41×10^{-26} J/T
Gyromagnetic ratio $\gamma/2\pi$	29.2 MHz/T	42.58 MHz/T

---

[1] For this reason, *spin* and *proton* are often used synonymously in MRI terminology.

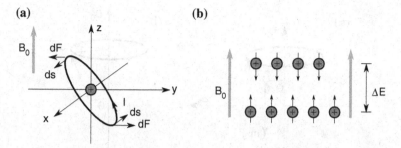

Fig. 5.1 a A current loop, such as a spinning proton, experiences forces that cause it to align with the direction of the external magnetic field B_0. **b** Depending on their spin orientation, the proton's magnetic moments μ (indicated by *arrows*) align either parallel or antiparallel to the magnetic field. A small energy difference ΔE in favor of the parallel orientation causes slightly more protons to assume the parallel orientation than the antiparallel orientation

as the spinning proton) with a current I flowing along a differential segment $d\vec{s}$ experiences a differential force $d\vec{F}$

$$d\vec{F} = I \; d\vec{s} \times \vec{B} \qquad (5.3)$$

that turns the current loop until it aligns with the magnetic field. The analogy is limited, because the spin orientation is a quantum effect, and spin orientations parallel and antiparallel to the magnetic field are possible. The parallel orientation has a slightly lower energy level than the antiparallel orientation, and more spins are oriented parallel than antiparallel to the magnetic field. The difference is called *spin excess*. The spin excess ΔN can be approximated by

$$\Delta N \approx N \frac{\hbar \omega_0}{2kT} \qquad (5.4)$$

where N is the total number of spins in the sample, $\hbar = h/2\pi$ is the reduced Planck constant, k is Boltzmann's constant, and T is the absolute temperature. ω_0 is the Larmor frequency that is introduced below. Typically, ΔN is a very small number in the order of less than ten spins per million. Since parallel and antiparallel spins cancel out, the net magnetic moment of a sample in an external field is dictated by the spin excess. We can multiply Eq. 5.4 with the magnetic moment of a proton and arrive at an approximation for the net magnetic moment M_0 of a sample in a magnetic field B_0:

$$M_0 = \frac{\rho_0 \gamma^2 \hbar^2}{4kT} B_0 \qquad (5.5)$$

where ρ_0 is the number of protons per unit volume and γ is the gyromagnetic ratio. Note that—by convention—the external magnetic field B_0 is always assumed to be oriented along the z-axis. The net magnetic moment M_0 in Eq. 5.5 is therefore

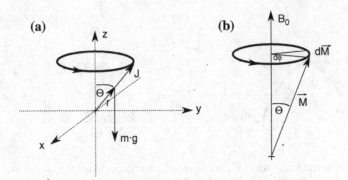

Fig. 5.2 a Spinning top analogy of precession. The vector $\vec{r}$ indicates the distance of the top's center of mass from the pivot point. Gravity ($m \cdot g$) acts on the center of mass in the direction of the negative z-axis. The spin axis $\vec{J}$, however, is at an angle Θ with respect to the z-axis, and the resulting differential change of the spin axis $\mathrm{d}\vec{J}$ lies in the x-y-plane tangential to the precession circle. **b** The analogy can be applied to the magnetic moments $\vec{M}$ that precesses at an angle Θ with respect to the magnetic field $\vec{B}_0$ and displaces the magnetic moment by $\mathrm{d}\vec{M}$ tangential to the precession circle (direction of $\vec{M} \times \vec{B}_0$)

also parallel to the z-axis and usually referred to as the *longitudinal magnetization*. Following the convention, we will denote the longitudinal magnetization M_z. The net magnetization is a tiny quantity and cannot be measured directly, particularly when it is superimposed over the very strong magnetic field B_0.

The spins do not align perfectly parallel to the magnetic field. Rather, they tend to precess around the z-axis similar to a toy top (Fig. 5.2). The spins obey the equation of motion,

$$\frac{\mathrm{d}\vec{M}}{\mathrm{d}t} = \gamma \vec{M} \times \vec{B}_0 \tag{5.6}$$

where γ is the gyromagnetic ratio. In accordance with Eq. 5.6, γ dictates the angular displacement of the magnetic moment $\vec{M}$ and therefore its precession speed. It can be shown that the precession frequency is proportional to both the gyromagnetic ratio and the magnetic field (Larmor equation):

$$\omega_0 = \gamma \cdot B_0 \tag{5.7}$$

where ω_0 is the precession frequency or *Larmor frequency*. For hydrogen, $\gamma/2\pi = 42.58\,\mathrm{MHz/T}$. Correspondingly, the precession frequency $f_0 = \omega_0/2\pi$ for hydrogen in a 1.5 T magnet is 63.87 MHz.

5.2 The Spin-Echo Experiment

We now define a new coordinate system (x', y', z) in which the (x', y')-plane rotates around the z-axis with the Larmor frequency (Fig. 5.3),

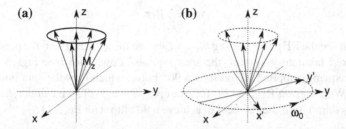

Fig. 5.3 a Multiple spins that are precessing around the z-axis create a net magnetization M_z in the z-direction. We do not see any magnetization in the x-y-plane, because the projections of the spins into the x-y-plane cancel out. **b** To better examine the transversal magnetization in the x-y-plane, we define a rotating coordinate system (x', y', z) where the x'-y'-plane rotates around the z-axis with the Larmor frequency ω_0 with respect to the fixed laboratory coordinate system. Relative to the rotating coordinate system, the spins do not precess and are fixed

$$\begin{bmatrix} x' \\ y' \\ z \end{bmatrix} = \begin{bmatrix} x \\ y \\ z \end{bmatrix} \begin{bmatrix} \cos \omega_0 t & \sin \omega_0 t & 0 \\ -\sin \omega_0 t & \cos \omega_0 t & 0 \\ 0 & 0 & 1 \end{bmatrix}. \tag{5.8}$$

The spins show precession with respect to the laboratory coordinate system but appear fixed when observed from the rotating (x', y', z) coordinate system. A radiofrequency (RF) signal can be constructed that is circularly polarized and whose magnetic component shows the time-dependent field strength

$$\vec{B}_{RF}(t) = B_1 \left(\hat{x} \cos \omega t + \hat{y} \sin \omega t \right) \tag{5.9}$$

where $\hat{x}$ and $\hat{y}$ are unit vectors along the x and y-axis, respectively. If ω is the Larmor frequency, the magnetic vector $\vec{B}_{RF}$ is stationary within the rotating coordinate system and therefore keeps stationary relative to the rotating magnetic moments. It is said that such a RF signal is *in resonance* with the spins, thus giving rise to the resonance part in magnetic *resonance* imaging.

Introducing radiofrequency energy in resonance can be compared to the one-dimensional case of a swing that is pushed in resonance, thus accumulating the energy to move with higher and higher amplitude. In the case of the RF pulse, the rotating B_1 field introduces a torque that attempts to tip the spins around the x'-axis with its own Larmor frequency $\omega_1 = \gamma \cdot B_1$. More precisely, the additional rotating RF field B_1 forces an additional change of the spin direction in the rotating (primed) coordinate system

$$\left(\frac{d\vec{M}}{dt} \right)' = \omega_1 \vec{M} \times \hat{x}' \tag{5.10}$$

In other words, the RF pulse *rotates* the spin around the x'-axis at a constant angular velocity. If the RF pulse is maintained only for a short duration τ, the angular change $\Delta \Theta$ can be quantified as

$$\Delta\Theta = \gamma B_1 \tau \tag{5.11}$$

We can see the RF pulse *flipping the spin* around the x'-axis. From the perspective of the fixed laboratory system, the spin "spirals" downward (see Figs. 5.4, 5.5). A very frequently applied RF pulse is a 90° pulse, which flips the spin into the x-y-plane. With a given RF field of, for example, $B_1 = 3\,\mu\text{T}$, we can determine the necessary duration of the RF field to achieve a 90° flip with Eq. 5.11:

$$\frac{\pi}{2} = 267.5 \times 10^6 \frac{\text{rad}}{\text{s}} \cdot 3 \times 10^{-6}\,\text{T} \cdot \tau \tag{5.12}$$

which yields $\tau \approx 2\,\text{ms}$. Note that we used γ in radians per second rather than the linear gyromagnetic ratio $\gamma/2\pi$ of 42.58 MHz.

In analogy to the swing that loses its accumulated resonant energy to friction and comes to a rest after some time, the spins, flipped under the influence of the RF electromagnetic pulse, lose their energy and return to the equilibrium state parallel to the z-axis. It is important to realize that the RF pulse, apart from flipping the spins, also induces *phase coherence*, that is, forces the spins to orient themselves closely parallel to the y'-axis. Immediately after the cessation of the RF pulse, therefore, the spins rotate in the x-y-plane *in coherence*, and the spins do not cancel each other out as they would in a random orientation.

When the spins are in phase coherence, their magnetization in the x-y-plane adds up to a resulting net *transversal magnetization* M_{xy}. The spin vector rotates with the Larmor frequency with respect to the fixed coordinate system but appears to be stationary with respect to the rotating (primed) coordinate system. Minor variations in the magnetic field (local field inhomogeneities) lead to small deviations of the precession frequency of different spins. The spins "fan out", a process called *dephasing*, and the resulting net transversal magnetization M_{xy} diminishes (Fig. 5.6). Loss of phase coherence is governed by a first-order differential equation,

$$\left(\frac{\mathrm{d}\vec{M}_{xy}}{\mathrm{d}t}\right)' = \frac{1}{T_2^*}\vec{M}'_{xy} \tag{5.13}$$

where $()'$ indicates a process with respect to the primed coordinate system. T_2^* is the decay constant, which depends on two components, a decay constant T_2 that is tissue-dependent and a decay constant T_2' that is dictated by field inhomogeneities such that

$$\frac{1}{T_2^*} = \frac{1}{T_2} + \frac{1}{T_2'} \tag{5.14}$$

The dephasing process is fast, with the decay constant T_2^* on the order of a few milliseconds. On a slower time scale, the spins lose more of their energy to the surrounding lattice, and as a consequence reorient themselves parallel to the z-axis. The longitudinal magnetization recovers and eventually reaches its equilibrium value as sketched in Fig. 5.3. Once again, the spin-lattice energy loss is governed

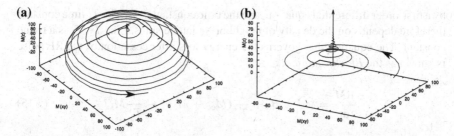

Fig. 5.4 Sketch of the total magnetization vector during RF-injection (**a**) and during recovery (**b**) with respect to the fixed coordinate system. In both cases, the apparent Larmor frequency has been reduced by about two orders of magnitude to better illustrate the spin rotation. During RF injection, the flip angle increases linearly with time (Eq. 5.11), and the total time for a 90° flip is assumed to be around 2 ms. Note that the spins can be flipped beyond 90°. Recovery follows two exponential functions with different time constants (Eqs. 5.16 and 5.17). Loss of transversal magnetization occurs rapidly, whereas the longitudinal recovery is relatively slow

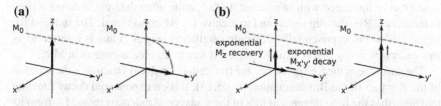

Fig. 5.5 Illustration of the spin orientation (see Fig. 5.4), but with respect to the rotating coordinate system. **a** The RF pulse at exactly the Larmor frequency linearly turns the spins around the x'-axis. After a RF duration of a few milliseconds, the spins have been rotated into the x-y-plane. Note that the spins can be rotated even further if the RF signal is applied even longer. **b** After cessation of the RF pulse, spins lose phase coherence (leading to decaying transversal magnetization M_{xy}) and more slowly return to the lowest-energy configuration parallel to B_0. Consequently, longitudinal magnetization M_z recovers

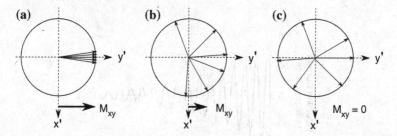

Fig. 5.6 Loss of phase coherence. **a** After application of the RF pulse, spins rotate in phase in the x-y-plane. With respect to the rotating primed coordinate system, they appear to be oriented closely around the y'-axis. **b** After cessation of the RF pulse, the spins lose their energy rapidly to neighboring spins, and their phase orientation randomizes. From within the rotating coordinate system, it appears as if some (*slower*) spins move backward, while some (*faster*) spins move forward within the x-y-plane. **c** The transversal magnetization M_{xy} diminishes as the spins dephase and eventually reaches an equilibrium value of zero

by a first-order differential equation, and the time constant is T_1. The T_1 time constant therefore depends on the density of neighboring lattice spins. Like T_2, T_1 is a tissue constant. The equation that governs the energy loss after cessation of the RF pulse is known as the *Bloch equation*:

$$\frac{d\vec{M}}{dt} = \gamma \vec{M} \times \vec{B}_0 + \frac{1}{T_1}(M_\infty - M_z)\,\hat{z} - \frac{1}{T_2^*}\vec{M}_{xy} \tag{5.15}$$

where M_∞ is the equilibrium longitudinal magnetization. The process of flipping the spins under the influence of the RF signal and the equilibration process (loss of phase coherence with loss of transversal magnetization) combined with the recovery of the longitudinal magnetization is illustrated in Fig. 5.4 for the fixed laboratory coordinate system and in Fig. 5.5 for the rotating (primed) coordinate system.

It is important to realize that the transversal magnetization M_{xy} rotates with the Larmor frequency. A rotating magnet gives off RF energy (induction!), and this RF signal can be measured with an antenna. In fact, quite often the same antenna is used to induce the RF spin-flip signal and to receive the RF echo signal. The induced RF signal follows an exponentially-decaying oscillation (Fig. 5.7) that is referred to as *free induction decay* (FID). This RF signal is key to all measurements in MRI.

The envelope of the RF signal, i.e., the free-induction decay function is the solution of the x-y-part of the Bloch equation (Eq. 5.13). It is an exponential decay function that describes the time-dependent loss of the transversal magnetization M_{xy} from its maximum at $t = 0$, M_0 (see Fig. 5.8):

$$M_{xy}(t) = M_0\, e^{-\frac{t}{T_2^*}} \tag{5.16}$$

Similarly, the solution of the Bloch equation for the longitudinal magnetization M_z is an exponential recovery function with the time constant T_1 as shown in Fig. 5.9:

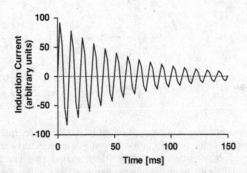

Fig. 5.7 Free-induction decay (FID). This graph shows qualitatively the current that is induced in the RF antenna coil by the spins rotating in the x-y-plane. As the spins lose their phase coherence, the net transversal magnetization decays and the FID current decays proportionally. For visualization purposes, the oscillation with ω_0 is not to scale

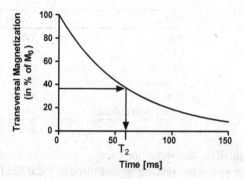

Fig. 5.8 Decay of the transversal magnetization with T_2^*. This function is also the envelope of the free-induction decay

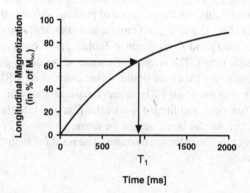

Fig. 5.9 Recovery of the longitudinal magnetization. After the cessation of the RF pulse, the spins are oriented in the x-y-plane, leading to a longitudinal magnetization of $M_z = 0$. As the spins lose their energy, they return to the equilibrium orientation parallel to the B_0 field. During this process, M_z recovers

$$M_z(t) = M_\infty \left(1 - e^{-\frac{t}{T_1}} \right) \tag{5.17}$$

After these derivations, we summarize the relaxation times:

- T_1: Recovery time constant for the longitudinal magnetization (parallel to B_0). Also referred to as spin-lattice relaxation, because the spins lose energy through thermal processes to the surrounding lattice.
- T_2: Decay time constant for the transversal magnetization (rotating moment perpendicular to B_0). Also referred to as spin–spin decay time, because dephasing occurs through energy loss from a spin to neighboring spins.
- T_2': Time constant for the loss of transversal magnetization caused by local field inhomogeneities, for example, local paramagnetic structures or charges. Unlike T_2 effects, field inhomogeneities can be canceled out by reversal of the spin direction. Unlike T_1 and T_2, T_2' is not dependent on the tissue characteristics.

Table 5.2 Decay constants T_1 and T_2 in some tissues at $B_0 = 1.5\,\text{T}$

Tissue	T_1 (ms)	T_2 (ms)
Fat	260	80
Liver	550	40
Muscle	870	45
White matter	780	90
Gray matter	900	100
Cerebrospinal fluid	2400	160

- T_2^*: Decay constant of the free induction decay. T_2^* reflects loss of transversal magnetization through spin–spin relaxation *and* through local field inhomogeneities (Eq. 5.14).

As a rule of thumb, $T_1 \gg T_2 > T_2^*$. Furthermore, fluids have a very long T_1 and T_2 (e.g., cerebrospinal fluid), but the presence of proteins shortens the time constants (e.g., blood). Fat has short T_1 and T_2, and muscle and other soft tissues lie in-between. Some typical values for T_1 and T_2 are given in Table 5.2; however, reported values for T_1 and T_2 span a wide range. The strong difference of the decay constants between soft tissues gives rise to the excellent tissue–tissue contrast of MRI.

We are presently able to obtain T_2^* by direct measurement: Obtain the spin-echo, measure its decay function, and fit an exponential function into the signal envelope to obtain T_2^*. However, we are interested in the tissue properties T_1 and T_2. We need to apply special tricks to obtain these constants. The need for such "tricks" gives rise to the pulse sequences.

5.3 The Spin-Echo Pulse Sequence

The term *pulse sequence* refers to the successive application of RF pulses and RF echo acquisition, and to the application of magnetic field gradients (covered in a later section). We need to recognize that the RF circuitry has to be switched from RF excitation to RF acquisition. This process can take a few milliseconds, in the process of which we lose relevant amplitude of the FID signal. It is more practical to wait a short time after the 90° pulse and allow the spins to slightly dephase, then apply a 180° pulse that flips the spins around the y'-axis back into the x-y-plane. After the 180° pulse, the spins appear mirrored along the y'-axis, but their relative speed is unchanged. Therefore, the spins now approach phase coherence (Fig. 5.10). Refocusing the spins is referred to as *spin-echo*.

We can now envision the following sequence of events that leads to the measurement of T_2^* in a small tissue sample:

1. Apply a 90° pulse to flip the spins into the x-y-plane. This pulse causes phase coherence, but the spins immediately begin to dephase.
2. Wait several milliseconds (let's call this time T_P) to allow some dephasing, then apply a 180° rephasing pulse. At this time, the spins begin to approach phase coherence.

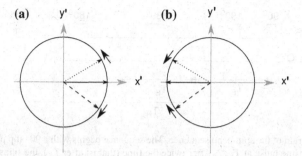

Fig. 5.10 Rephasing after application of an 180° pulse. **a** Initially, the spins are dephasing with a decay constant of T_2^*. Three representative spins are shown: One (solid) that precesses with the Larmor frequency, one (dotted) that lags behind, and one (dashed) that speeds ahead. **b** With the 180° rephasing pulse, the spins are flipped around the y'-axis. Their relative speed is unchanged, but the faster speed of the dashed spin and the slower speed of the dotted spin causes them to converge towards phase coherence

3. Wait another few milliseconds (T_P, again) and acquire the FID signal. Find the signal maximum and its envelope and obtain T_2^*. Let us denote the time from the start of the sequence to the echo acquisition T_E so that $T_E = 2T_P$.

5.3.1 Measurement of T_2

Although T_2^* is a relevant parameter in some special applications, a more important tissue parameter is T_2, and its determination a separate goal. Interestingly, the field inhomogeneity effects (T_2') cancel out as the spins reverse their direction, but the thermodynamic (spin–spin relaxation) effects do not. We can repeat the application of the 180° rephasing pulse several times, and each time, the FID peak is lower due to T_2 losses only. More precisely, subsequent FID peaks decay with T_2, and the analysis of repeated echoes acquired with repeated rephasing pulses yields the tissue constant T_2. The full spin-echo sequence, therefore, starts with a 90° pulse followed by a 180° rephasing pulse at $T_E/2$ and acquisition of the first FID amplitude at T_E. With the same time intervals, 180° rephasing and RF acquisition are repeated until a sufficient number of echoes have been acquired to obtain T_2 by analysis of the peak decay (Fig. 5.11).

5.3.2 Measurement of T_1 Through Incomplete Recovery

We were able to measure the transversal magnetization, because the rotating spins emit a RF signal whose amplitude is proportional to M_{xy}. The longitudinal magnetization cannot be directly accessed in the same fashion, and T_1 cannot be determined directly.

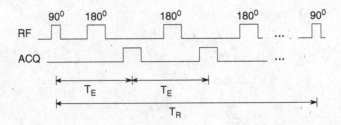

Fig. 5.11 Diagram of the spin-echo sequence. The sequence begins with a 90° flip pulse followed by a 180° rephasing pulse at $T_E/2$. After twice the time (that is, after T_E), the spins regain phase coherence and the FID signal is acquired. The application of the 180° rephasing pulse and signal acquisition is repeated several times. After some time—T_R—the sequence can be repeated

It is possible, however, to access T_1 indirectly by not allowing the longitudinal magnetization to fully recover after a 90° flip. If we choose a short repetition time T_R (Fig. 5.11), more specifically, a repetition time that is near the longitudinal relaxation constant T_1, the magnetization M_z does not have enough time to reach its equilibrium value. *However, recovery is faster in tissues with shorter T_1.* After identical repeat times, therefore, the tissue with the shorter T_1 will have a larger longitudinal magnetization than the tissue with longer T_1. On the next application of the 90° pulse, the *actual M_z* is flipped, and the subsequent spin echo amplitude is therefore proportional to M_z at the time T_R. Clearly, the tissue with shorter T_1 now provides a larger FID signal than the tissue with longer T_1.

When incomplete recovery is chosen through short T_R, decay effects with T_2^* still play a role. To minimize T_2^*-effects, a short T_E is chosen in addition to the short T_R to keep the transversal decay small.

5.3.3 Measurement of Proton Density

We can take the idea from the previous section one step further. If we allow full recovery of M_z by choosing a long T_R, but prevent T_2^*-effects from influencing our signal by choosing a short T_E, the main determinant of the FID amplitude is the number of aligned spins. In other words, the FID amplitude is proportional to the hydrogen content (i.e., the water content) of the tissue. This tissue parameter is referred to as *proton density* or PD.

5.3.4 The Significance of T_E and T_R

We have seen in the previous sections that the spin-echo sequence allows us to determine T_1, T_2 or the proton density. This versatility makes the spin-echo sequence

Table 5.3 How to obtain T_1, T_2, or PD (proton density) contrast with appropriate choices of T_R and T_E

Desired contrast	Choice of T_E	Choice of T_R
T_1	short	short
T_2	long	long
PD	short	long

one of the most widely used sequences. In summary, the radiologist *chooses* the measurement times T_E and T_R, and the amplitude of the FID signal carries information on proton density, T_1, or T_2 following Table 5.3. More specifically, the FID amplitude maximum A_{max} is governed by Eq. 5.18:

$$A_{max} \propto \rho e^{-\frac{T_E}{T_2}} \left(1 - e^{-\frac{T_R}{T_1}}\right) \tag{5.18}$$

We can see from Table 5.3 that a fourth option exists, that is, choosing a long T_E in combination with a short T_R. This is an undesirable combination, because it mixes T_1 and T_2 effects, and it allows M_{xy} to decay unnecessarily.

How suitable choices for T_E and T_R relate to T_1 and T_2 is further illustrated in Figs. 5.12 and 5.13. With a short T_E, little decay of the transversal magnetization occurs. With a very long T_E, the transversal magnetization decays strongly, irrespective of the tissue. Two different tissues with different T_2 yield good contrast if T_E is chosen near the tissue's T_2 decay constants. Note the different behavior of T_1 and T_2 contrast, provided that the amplitude in Eq. 5.18 is used as image intensity: In a T_2-weighted image, the transversal magnetization decays faster when T_2 is short, and tissues with short T_2 appear darker in the image than tissues with long T_2. Conversely, tissues with a long T_1 appear darker in a T_1-weighted image than tissues with a short T_1, because the longitudinal magnetization recovers faster in tissues with short T_1, and more transversal magnetization is available at the next 90° flip.

5.4 From NMR to MRI: The Gradient Fields

Up to this point, we discussed spin effects in a small, homogeneous sample. Magnetic resonance *imaging* requires that the spin-echo signal is spatially resolved and carries local information about inhomogeneous tissue. The FID signal amplitude A can then be displayed as a function of their spatial location, that is, $A(x, y, z)$, and viewed as an image. Recall that RF excitation can only occur at resonance. An RF signal that is not in resonance does not cause the spins to flip or achieve phase coherence. The secret to resolving the signal spatially is to modulate the Larmor frequency along one spatial axis and exploit the spatially-dependent resonance frequency. Recall

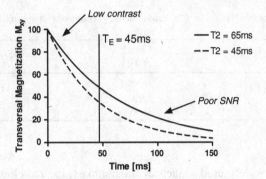

Fig. 5.12 Significance of T_E in obtaining T_2 contrast. We assume that T_R is long enough for M_z to fully recover, that is, $T_R \gg T_1$. When T_R is long enough to allow M_z recovery, T_1-effects no longer play a role. The transversal magnetization M_{xy} decays with T_2, and the tissue with the shorter T_2 (*dashed line*) has a faster decaying signal. When a measurement of the FID amplitude is taken at T_E, the tissue with the shorter T_2 has a lower signal strength and therefore a *darker shade* in the image than the tissue with longer T_2. Ideally, T_E is selected long enough to obtain a strong difference in M_{xy} between the tissues, but short enough to avoid unnecessary signal decay with its associated poor SNR

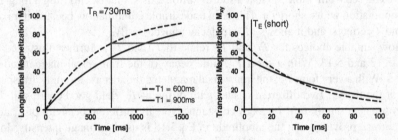

Fig. 5.13 Significance of T_E and T_R in obtaining T_1 contrast. The pulse sequence is repeated rapidly, and unlike in the situation in Fig. 5.12, the longitudinal magnetization M_z is *not* allowed to fully recover by choosing a short T_R (*left*). At T_R, a 90° flip pulse is applied (*arrows*), and the longitudinal magnetization present at $t = T_R$ is converted into transversal magnetization. Thus, a tissue with a short T_1 (*dashed line*) and consequently faster recovery of M_z has a larger transversal magnetization M_{xy} *immediately after* the 90° flip. Now, the FID is measured with very short $T_E \ll T_2$ to avoid T_2-related magnetization decays (*right*). In this case, the tissue with the shorter T_1 and faster recovery of M_z has a larger signal and consequently brighter image value than the tissue with longer T_1. Ideally, T_R is selected near one of the T_1, and T_E is as short as technically possible

the Larmor equation, $\omega_0 = \gamma \cdot B_0$. If we superimpose a spatially-varying magnetic field (the *gradient field*) $B_G(s)$, the Larmor frequency becomes dependent on the location s:

$$\omega_0(s) = \gamma \cdot B(s) = \gamma \cdot (B_0 + B_G(s)) \tag{5.19}$$

Here, s is some vector. Usually, s coincides with one axis of the laboratory coordinate system. In the following sections, we will use B_0 as the static primary magnetic field (assumed to be completely homogeneous) and B_G as the superimposed spatially dependent gradient field. Both magnetic fields are parallel to the z-axis, and we can interpret the magnetic fields as scalars. $\vec{G}_s$ is the field gradient,

$$\vec{G}_s = \nabla \cdot (B_0 + B_G(s)) = \nabla \cdot B_G(s) \qquad (5.20)$$

where ∇ is the three-dimensional gradient operator $(\partial/\partial x, \partial/\partial y, \partial/\partial z)$. Usually, the gradient follows one axis x, y, or z of the laboratory coordinate system (although gradients along multiple main axes could be superimposed to obtain an off-axis gradient). Moreover, the gradient is linear within the field of view (in other words, $\vec{G}_s$ is constant), and the combined field at a location $\vec{s}$ is calculated as

$$B(\vec{s}) = B_0 + \vec{s} \cdot \vec{G}_s \qquad (5.21)$$

5.4.1 The Slice Encode Gradient

Consider the arrangement in Fig. 5.14. A gradient field $B_z = G_z \cdot z$ is superimposed over the primary magnetic field such that the Larmor frequency becomes dependent on the z coordinate:

$$\bar{\omega}_0(z) = \gamma \cdot (B_0 + B_z(z)) = \gamma \cdot (B_0 + z \cdot G_z) \qquad (5.22)$$

Any NMR phenomenon can only take place in resonance when the RF signal matches exactly the Larmor frequency. When the z-gradient is applied, only spins in the section very close to the x-y-plane can be manipulated when the RF signal has the same frequency as the original Larmor frequency when $G_z = 0$. Consequently, any spin-echo that is received *must* originate from this thin slice, because any section above or below this slice has not achieved phase coherence and cannot emit an echo.

Let us assume $B_0 = 1.5$ T for an example. We furthermore apply a gradient $G_z = 10$ mT/m. Also recall that the gyromagnetic ratio for hydrogen is $\gamma/2\pi = 42.58$ MHz/T. The RF frequency for the slice at $z = 0$ is therefore 63.87 MHz. However, if we tune the RF generator to produce 63.891 MHz, the slice parallel to the x-y-plane at $z = 5$ cm is in resonance, and the 90 and 180° pulses only affect this slice. *By choosing a RF frequency that deviates from the original no-gradient Larmor frequency, we can choose the location of the slice to be excited.* For this reason, a gradient that is activated during the RF excitation phase is called the *slice select gradient* or slice encode gradient. A volumetric MR image is obtained by acquiring one cross-sectional slice and then moving to a subsequent slice by repeating the acquisition with a different RF frequency. Staying with the example above, we can advance in 10 mm steps in the z-direction by increasing the injected RF frequency by 4258 Hz between slice acquisitions.

How thick is the slice that is excited by the RF signal? Deviations of a few ppm (parts per million) from the Larmor frequency are sufficient to no longer meet the resonance condition. However, the boundary of the excited slice is not abrupt. Rather,

82

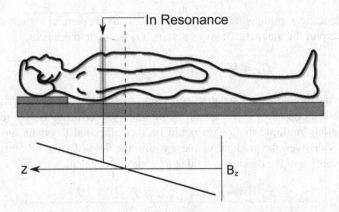

Fig. 5.14 Selective excitation of a slice parallel to the x-y-plane. A field gradient is superimposed over the B_0 field so that the field strength and therefore the Larmor frequency increases linearly in the z-direction. Since proton spin and phase can only be influenced by an RF signal *in resonance*, only a small slice of tissue experiences the 90° spin flip and the subsequent phase coherence. Any spin-echo that is measured is therefore *known* to originate from this thin slice of tissue. If no gradient field were applied ($B_z = 0$), the entire volume would be excited by the RF signal. If the gradient is applied, the original Larmor frequency excites the slice at $z = 0$ (*dashed line*). However, by slightly increasing the RF frequency, a slice in the positive z direction is excited (*gray shaded area, arrow*), and the choice of the RF frequency allows us to select the location of the excited slice with respect to the z-coordinate

some phase coherence is achieved a small distance from the center of the plane. The RF signal can be designed to have a specified *bandwidth*, that is, its Fourier transform is approximately constant from $\omega_0 \pm \Delta\omega$ and then rapidly drops to zero. It can be shown that the *sinc* signal, that is,

$$f(t) = \text{sinc}(2\pi\,\Delta f t) \tag{5.23}$$

has a nonzero Fourier transform between $-\Delta f$ and $+\Delta f$, and rapidly drops to zero for higher frequencies ("boxcar" function, see Sect. 1.4). By multiplying the sinc-function with the unmodulated RF signal of frequency ω_0, the center of the boxcar function is shifted to $\pm\omega_0$, and the positive half of the Fourier transform of the modulated RF signal becomes approximately

$$F(\omega) = \begin{cases} 0 & \text{for} \quad \omega < \omega_0 - \Delta\omega \\ F_0 & \text{for} \quad \omega_0 - \Delta\omega \leq \omega \leq \omega_0 + \Delta\omega \\ 0 & \text{for} \quad \omega > \omega_0 + \Delta\omega \end{cases} \tag{5.24}$$

Here, $\Delta f = \Delta\omega/2\pi$ is the bandwidth of the sinc-modulated RF signal (Fig. 5.15). The duration of the signal, more precisely, the number of sidelobes, determines the final bandwidth. In fact, the number of zero crossings, n_z, of the RF pulse is an important quantity in the design of a RF pulse. The zeros of $\text{sinc}(x) = \sin(x)/x$ are all nonzero integer multiples of π. The number of zero crossings is related to the

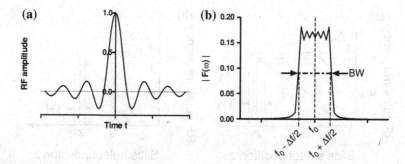

Fig. 5.15 a Modulation profile (envelope) for a RF signal that approximates a *sinc* pulse or $\sin(\omega t)/\omega t$. The theoretical sinc-pulse has infinite support, but a practical RF pulse is time-limited, in this example to 5 cycles. **b** Its Fourier transform approximates a "boxcar" function with a bandwidth BW $= \Delta f$ around the center frequency f_0

bandwidth through Eq. 5.25,

$$n_z = [\Delta f \cdot \tau_{RF}] \qquad (5.25)$$

where Δf is identical to the bandwidth BW, τ_{RF} is the total duration of the RF pulse, and the square brackets [] indicate the *floor* function.[2] The bandwidth can be seen as the frequency range in which frequency components exist. For this reason, the bandwidth directly and linearly influences the thickness of the slice excited by the *sinc* RF pulse.

From the above considerations and Fig. 5.16, both the gradient strength and the RF bandwidth can be used to choose the slice thickness. A narrow bandwidth improves the signal-to-noise ratio:

$$\text{SNR} \propto \frac{1}{\sqrt{\Delta f}} \qquad (5.26)$$

However, a narrow bandwidth causes other undesirable effects, such as chemical shifts.[3] The choice of the bandwidth is therefore a balance between different artifacts.

The slice encode gradient is applied during RF excitation. Since the Larmor frequency slightly differs over the thickness of the slice, the spins are not in phase. Rather, spins accumulate phase in a way that bears much similarity to the phase encoding we need for the encoding in the third dimension (see Sect. 5.4.4 below). For this reason, a second gradient pulse is applied right after the first pulse of the slice encode gradient, but with opposite sign and with half the duration. This inverted

---

[2] The `floor()` function in the C language indicates the largest integer less than or equal to its argument.

[3] Chemical shifts are very small shifts of the Larmor frequency depending on the environment, for example, whether the proton (i.e., hydrogen) is bound in a water or a lipid molecule. Chemical shifts are the basis of NMR analysis, but play a minor role in MRI. We will not further discuss chemical shifts in this book.

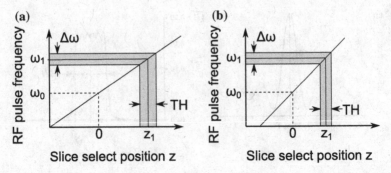

Fig. 5.16 Both the bandwidth of the RF signal ($\Delta\omega = 2\pi\Delta f$, see Fig. 5.15) and the gradient strength can influence the selected slice thickness. The diagonal line represents the gradient $\gamma \cdot (B_0 + z \cdot G_z)$. In Panel **b**, the gradient G_z is larger than in Panel **a**. The no-gradient Larmor frequency ω_0 selects a slice around $z = 0$, indicated by the dashed line. The slice thickness TH can be influenced by the RF bandwidth $\Delta\omega$, and a larger bandwidth excites a thicker slice centered on z_1, where $\omega_1 = \gamma \cdot (B_0 + z_1 \cdot G_z)$. When G_z is larger (**b**), the slice thickness is smaller

gradient is referred to as the *rephasing gradient*, and it can be seen as "rewinding" the phase. After application of the rephasing gradient, all spins have zero phase.

5.4.2 Fourier-Encoding with the Gradient

The Larmor frequency changes *instantly* with the magnetic field strength. Once the slice select gradient has been turned off, all excited spins return to their original, non-gradient Larmor frequency $\omega_0 = \gamma B_0$. Let us now introduce a second gradient $\vec{G}$ along a direction $\vec{r}$ in the x-y-plane, whereby $\vec{r}$ subtends an angle φ with the x-axis. The gradient, superimposed over the primary magnetic field, creates a direction-dependent Larmor frequency:

$$\omega(r) = \gamma \cdot (B_0 + \vec{r} \cdot \vec{G}) \tag{5.27}$$

It is justifiable to interpret the measurable magnetization with the gradient $\vec{G}$ as a projection, because each frequency component contains the cumulative spin-echoes from thin "strips" orthogonal to the gradient direction. Let us denote the new axis that is perpendicular to r as s. Assuming that we excited only a thin slice by making use of the slice select gradient, these strips would be parallel to the s-direction, and we can write the cumulative transversal magnetization $P(r, \varphi)$ (i.e., the projection onto the r-axis) as

$$P(r, \varphi) = \int M_\perp(x, y)\, ds \tag{5.28}$$

where $M_\perp(x, y)$ is the transversal magnetization of an infinitesimal volume at the coordinates x and y. It is important to realize that the *measured signal* is a harmonic oscillation. In fact, the demodulated signal with respect to the rotating coordinate system is a time-dependent complex function $m_\perp$,

$$m_\perp(t, \varphi) = \int P(r, \varphi)e^{-j\gamma Grt}\, dr = \int P(r, \varphi)e^{-jkr}\, dr \qquad (5.29)$$

where we introduce a normalized time k defined as $k = \gamma Gt$. The most interesting feature of Eq. 5.29 is its Fourier-transform character. In fact, the measured signal $m_\perp(t)$ and the projection $P(r)$ for any angle φ form a Fourier transform pair so that the projection can be recovered from the signal by an inverse Fourier transform,

$$P(r, \varphi) = \frac{1}{2\pi} \int m_\perp(k, \varphi)e^{+jkr}\, dk \qquad (5.30)$$

We can claim that the projection is *Fourier-encoded* in the measured signal by the magnetic gradient. The relationship to the Radon transform in computed tomography is interesting to note. In fact, the projections could be used to reconstruct the cross-sectional image by filtered backprojection, but in practice, a Fourier matrix, called *k-space matrix* with two orthogonal normalized time axes k_x and k_y can be filled by Cartesian sampling.

5.4.3 The Frequency Encode Gradient

In a simplification of the considerations from the previous section, we can instead introduce a second gradient, orthogonal to the slice select gradient, that is applied *during the RF echo acquisition*. Let us denote this gradient G_y, whereby the resulting magnetic field with the frequency encode gradient enabled becomes:

$$B(y) = B_0 + y \cdot G_y \qquad (5.31)$$

At this point, the frequency of the received RF signal depends on the y-position of the spins (Fig. 5.17). With the Fourier transform, we can now determine the signal strength as a function of the frequency, and consequently as a function of the y-position. We have localized the signal in the z direction by applying the gradient G_z during RF injection, and we have localized the signal in the y direction by applying the gradient G_y during echo acquisition.

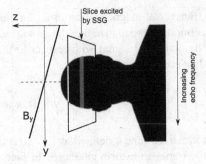

Fig. 5.17 The Larmor frequency of the spins reacts instantaneously to changes in the magnetic field. If a gradient (for example, in the y-direction) is turned on during echo acquisition, the frequencies of the spins are proportional to their position along the y-axis. Quadrature demodulation and subsequent Fourier analysis yields the signal strength as a function of the y-coordinate

5.4.4 The Phase Encode Gradient

With the slice select and frequency encode gradients, we have a one-dimensional signal from a two-dimensional slice of tissue, and one spatial direction is encoded in the frequency. Similar to Fourier-based reconstruction in CT, we need to fill an empty Fourier-domain spaceholder with data to obtain the cross-sectional image by inverse Fourier transform. With the definition of the normalized time $k = \gamma G t$ (Sect. 5.4.2), the Fourier-domain spaceholder is called *k-space matrix*.

The solution to the defined filling of the k-space matrix is to make use of the gradient Fourier encoding introduced in Sect. 5.4.2. A gradient G changes the Larmor frequency along an arbitrary direction r in the x-y-plane. If the gradient is applied for only a short time τ, the spins accumulate a spatially-dependent phase ϕ:

Fig. 5.18 Filling of the k-space matrix with repeated acquisitions at different gradient strengths G_x. Each gradient defines a position along k_x (*arrow*), and a subsequent frequency-encoded acquisition fills one column of the k-space matrix at k_x (*gray shaded area*). With repeated acquisitions that use different strengths for G_x, the entire k-space matrix can be filled

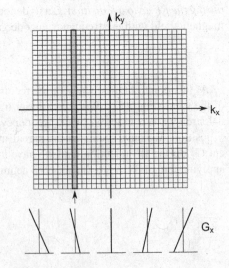

$$\phi(r) = \int\limits_0^\tau r \cdot \omega_G(r, t) \, \mathrm{d}t = \gamma \cdot r \cdot \int\limits_0^\tau G(t) \, \mathrm{d}t. \qquad (5.32)$$

Equation 5.32 leads to the same type of Fourier encoding of the demodulated signal by the gradient *through its phase*. Since we have restricted RF excitation to the x-y-plane and we are decoding the signal along the y-axis with a gradient in the y-direction, the phase encode gradient needs to be orthogonal to the two other gradients, that is, in our example in the x-direction. Whereas the frequency encode gradient provides us with a complete column of data in the k-space matrix, the phase encoding provides us with a single point along the k_x axis. We can therefore interpret the choice of one specific gradient G_x as positioning of the acquired row along k_y on a specific position k_x (Fig. 5.18). To completely fill the k-space matrix, repeated acquisitions need to be performed with different gradient strengths G_x (and its associated position k_x) until all rows of the k-space matrix are filled and the desired cross-sectional image emerges as the inverse Fourier transform of the k-space matrix.

5.5 Putting Everything Together: Spatially-Resolved Spin-Echo Acquisition

We have now assembled all components for magnetic resonance imaging, and a summary of how the individual components fit into the bigger picture is helpful. We therefore revisit the spin-echo sequence, but with the gradients included (Fig. 5.19). The sequence follows the basic structure of the spin-echo experiment (Fig. 5.11) introduced earlier. In addition, a slice select gradient restricts RF excitation to a defined slice, and two orthogonal gradients allow Fourier encoding of the signal. Since one acquisition fills one column of the k-space matrix, the sequence needs to be repeated with a different phase encode gradient for each column of the k-space matrix. The significance of T_E and T_R to obtain T_2-, T_1- or proton density-weighted images remains the same as in Sect. 5.3.

The necessity to repeat the sequence multiple times gives rise to the question of exam duration. Two key determinants are T_R and the number of applications of G_x to fill the k-space matrix. If this number is denoted n_x, the total acquisition time for one slice is

$$t_{\text{Slice}} = T_R \cdot n_x \cdot n_{\text{ex}} \qquad (5.33)$$

where n_{ex} is the number of repeated acquisitions with the *same* gradient. Averaging multiple acquisitions improves the signal-to-noise ratio by

$$\text{SNR} \propto \sqrt{n_{\text{ex}}} \qquad (5.34)$$

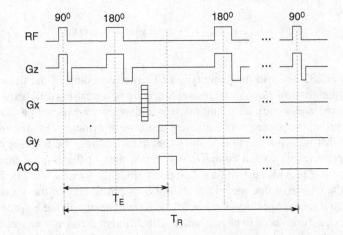

Fig. 5.19 Spin-echo sequence to obtain a cross-sectional image. The basic structure follows the spin-echo experiment (Fig. 5.11), but with the added spatial gradients. Any RF injection needs to be accompanied by the slice select gradient G_z to restrict spin excitation to a small plane. Positioning in k-space along k_x is performed with the phase encode gradient G_x that is applied between the rephasing pulse and the echo acquisition. A frequency encode gradient G_y is applied during echo acquisition to obtain the Fourier encoded information along the y-axis

For example, a T_1-weighted image sequence with a resolution of 256×256 pixels and a relatively short T_R of 500 ms requires roughly 2 minutes per slice. A T_2-weighted image, which requires a longer T_R of 2 s (and assuming $n_{ex} = 4$) would take more than half an hour per slice. Although the spin-echo sequence is very popular due to its versatility, faster sequences are available that significantly shorten the acquisition time, although often at the expense of image quality.

5.6 Other Imaging Sequences

Many alternative sequences exist, some of them with the primary purpose to reduce the acquisition time. As one simple example, a sequence could make use of the Hermitian symmetry of the Fourier space and merely fill half of the k-space matrix, thus reducing acquisition time by a factor of two, but at the expense of lower SNR. This sequence is known as Half-fourier Acquisition Single shot Turbo spin-echo ("HASTE").[4] In another example, called *fast spin-echo*, FSE, repeated 180° pulses are used to rephase the spins and elicit a spin-echo after one initial 90° excitation pulse. By applying a phase encode gradient after each 180° rephasing pulse, multiple lines of the k-space matrix can be filled with one 90° excitation (and thus within T_R). However, since the echo decays over time, a 90° excitation needs to be performed

---

[4] It seems to be a convention in MRI to often find cute or funny acronyms for the sequences. Enjoy.

every few lines. Typically, FSE is used for T_2-weighted images, and the acquisition is shortened by a factor identical to the number of k-space lines filled between 90° pulses.

We will introduce a few key sequences in this section and explain their fundamental purpose. The goal of this section is more to provide an idea of what different sequences can achieve rather than provide a comprehensive list of available sequences.

5.6.1 Gradient-Recalled Echo Sequences

In Sect. 5.4.1, we mentioned that the slice encode gradient is followed by a negative gradient lobe of half the area, whose purpose is the rephasing of the spins. After a RF pulse with slice encode gradient and rephasing lobe, the spins are reset to have a 0° phase. Analogously, beginning the frequency encode gradient with a negative dephasing lobe, followed by the regular frequency encode gradient leads to phase coherence very similar to the spin-echo elicited by the 180° rephasing RF pulse. In this fashion, the FID echo signal can be obtained through the application of a gradient *instead* of the 180° rephasing pulse. Such a sequence is called *gradient recalled echo* or GRE sequence. A typical GRE sequence diagram is shown in Fig. 5.20.

GRE sequences have two advantages over the spin-echo sequence. First, because of the absence of the 180° rephasing pulse, T_E can be shortened, and second, it is not necessary to flip the spins by 90°. Recovery from a lower flip angle (see next section) is faster, and T_R can be shortened. Both factors allow for faster acquisition times. Depending on T_E and T_R, GRE sequences can provide T_1-, T_2^*-, and proton density-weighted images, although it is most often used to acquire T_1-weighted images because of its shorter T_R and consequently overall shorter acquisition time.

Further speed improvement is possible by adding a spoiler gradient. We discussed before that field inhomogeneities shorten T_2^* and T_1. After echo acquisition, a strong gradient pulse (spoiler pulse) can be applied that "swirls" the protons and shortens the time to equilibrium. Further reduction of T_R is possible, notably in T_2^*-weighted sequences. GRE sequences are not typically used for true T_2-weighted images, because the gradient recalled echo does not accommodate spin inversion in a similar manner as the repeated 180° pulse in a spin-echo sequence does.

5.6.1.1 Fast, Low-Angle Shot

The FLASH sequence is a good example of a high-speed GRE sequence. For a FLASH sequence, the initial 90° pulse is shortened to flip the spins halfway between their equilibrium position and the x-y-plane. A transversal magnetization exists as the projection of the spins onto the x-y-plane, and the projected transversal magnetization is smaller than it would be with a full 90° flip. Since phase coherence is achieved, a RF signal can be received, but the signal is attenuated by a factor of $\sin \alpha$ over the conventional spin-echo sequence, where α is the flip angle. Corresponding with the

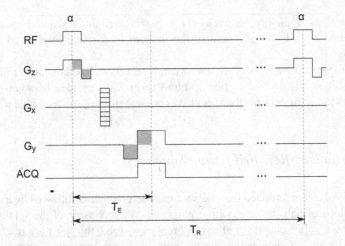

Fig. 5.20 Gradient recalled echo (GRE) sequence. In a GRE sequence, the initial RF pulse may flip the spins by an angle $\alpha < 90°$, and no $180°$ rephasing pulse is applied. Instead, the FEG pulse is used to achieve phase coherence after an initial dephasing lobe. Acquisition takes place during the phase coherence period. Note that the shaded regions under the SSG and FEG pulses have the same area

lower signal, SNR is lower, and the images appear more noisy. However, a simple relationship between SNR and flip angle does not exist. Rather, a preferred flip angle exists that maximizes SNR for very short T_R, because the longitudinal magnetization is kept near its equilibrium. The flip angle that optimizes SNR is known as *Ernst angle* α_{opt}, for which holds:

$$\cos \alpha_{\mathrm{opt}} = e^{-\frac{T_R}{T_1}} \tag{5.35}$$

Since the spins are close to their equilibrium position, T_R can be shortened considerably. A gradient is applied to spoil the transverse magnetization, and very short T_E becomes possible. The FLASH sequence is often used to acquire 3D MR images with T_1 weighting, when a multi-slice acquisition would take a very long time with the spin-echo sequence. Since T_R can be shortened to tens of milliseconds, a 3D volume scan of the head could be completed in a few minutes. With the very fast acquisition times possible with GRE sequences such as FLASH, motion artifacts are reduced, and it even becomes possible to image the uptake of contrast agents. In conjunction with gated acquisition (i.e., acquisition synchronized with the heartbeat), the heart can be imaged.

5.6.2 Inversion Recovery Sequence

The inversion recovery sequence is a sequence to achieve excellent T_1 contrast. Its main disadvantage is that it requires a very long T_R. Unlike the spin-echo sequence, the inversion recovery sequence begins with a $180°$ flip, which flips the longitudinal magnetization from M_z to $-M_z$ (inversion). Recovery requires at some point—depending on the tissue's T_1—to go through $M_z = 0$. After a recovery period T_I, a 90° pulse is applied that flips the current M_z into the x-y-plane, and a transversal magnetization is obtained that is proportional to M_z at the time T_I. With a conventional rephasing pulse or gradient, the FID echo can be measured at the echo time T_E. Note that T_I is another variable time parameter, in addition to T_R and T_E. However, T_E is generally kept as short as possible to eliminate T_2-effects and T_R is kept long to allow sufficient recovery (unlike T_R in the T_1-weighted spin-echo sequence, where it is kept short to force incomplete recovery). Since recovery occurs from the highest possible energy level (the point farthest away from equilibrium), recovery times of 5 s or longer are not uncommon.

The demodulated signal intensity A_{max} can be approximated as (cf. Eq. 5.18):

$$A_{max} \propto \rho \left(1 - 2e^{-\frac{T_I}{T_1}} + e^{-\frac{T_R}{T_1}} \right) \tag{5.36}$$

and we can see from Eq. 5.36 that no terms with T_2 occur. Note also that some choices of T_I and T_R may lead to negative values of A_{max}. Intuitively, this happens when T_I is so short that the spins are still flipped below the x-y-plane. In fact, we can solve Eq. 5.36 for $A_{max} = 0$ and find for $T_R \gg T_I$ that the signal disappears for $T_I \approx T_1 \cdot \ln 2$. this phenomenon of suppressing the signal from specific tissues gives rise to a set of tissue-attenuating sequences of which two are listed below.

5.6.2.1 Short-Tau Inversion Recovery

With a short inversion recovery time T_I, adipose tissue with its short T_1 is attenuated or even completely suppressed. Typical values for a STIR sequence are $T_I = 140$–180 ms and $T_R = 2500$ ms. STIR images can sometimes emphasize tissue detail where fat content can lead to reduced contrast (this includes the brain), and it can reduce the distracting chemical shift that is associated with hydrogen embedded in fat.

5.6.2.2 Fluid Attenuated Inversion Recovery

The opposite to STIR is FLAIR with a long T_I to suppress the signals from fluids, such as CSF. Since fluids have a high proton density, their signal can be overwhelming, and FLAIR images can bring out anatomical detail that are otherwise overshadowed by the fluid signal. A typical FLAIR sequence uses $T_I = 2400$ ms and $T_R = 7000$ ms or longer. The very long T_R is required for full M_z recovery. With such long T_R, FLAIR sequences become extremely time-consuming and are rarely performed.

5.6.3 Echo Planar Imaging

The purpose of the echo planar imaging (EPI) sequence is to achieve an extremely short acquisition time. In fact, it is possible to fill the entire k-space matrix within one T_R-period. The EPI sequence starts with a 90° spin-flip pulse followed by a strong combined PEG and FEG gradient to position the spins in one corner of the k-space matrix. A 180° rephasing pulse follows immediately. After the rephasing pulse, an oscillating frequency encode gradient alternates the direction in which the lines of the k-space matrix are filled per echo, whereby a simultaneous "blip" of PEG advances the readout by one line in the k-space matrix (Fig. 5.21). Therefore, the EPI sequence fills the k-space matrix in a zig-zag pattern. FID amplitude is uneven, and a maximum FID signal occurs at the effective echo time T_E, which is not an adjustable parameter in EPI. Due to T_2^*-effects, acquisition of the entire k-space matrix must occur within a period less than T_2^* (around 20–50 ms), which places a high demand on the gradient system and the RF acquisition rate. A gradient-recalled alternative exists, which does not need the 180° refocusing pulse.

EPI images typically have a low resolution (64×64 or 64×128) with very poor SNR. The EPI technique is particularly susceptible to artifacts, such as chemical shifts and local magnetic inhomogeneities, but it is one of the few techniques with real-time capabilities, being capable of delivering images with frame rates of 20 fps or higher. EPI has been used for monitoring physiological processes, such as the beating heart.

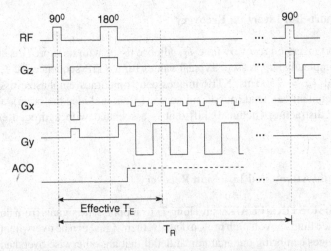

Fig. 5.21 Spin-echo planar imaging (EPI) sequence. EPI begins similar to the spin-echo sequence, but there is a strong gradient between the 90° flip pulse and the 180° rephasing pulse, which positions the spins at the corner of the k-space matrix. With alternating FEG pulses and short PEG blips, the k-space matrix is filled in a *zig-zag* pattern within one T_R-period

5.7 Technical Realization

The core of a MRI device contains three key components: The primary B_0 magnet coil, the gradient subsystem, and the RF subsystem. In addition, a computer for timing and sequence control and for data processing is required. An overview of an entire MRI system is shown in Fig. 5.22, and some of the key components are discussed below.

5.7.1 B_0 Magnet

In early MRI devices, the B_0 field was generated by permanent magnets, which were able to provide up to 0.2 T. However, a higher strength primary field has a number of advantages, predominantly shorter recovery times (shorter T_1 and thus shorter T_R), and a better SNR (cf. Eq. 5.4). Higher magnetic field strengths require electromagnets, but conventional coils are impractical due to thermal losses. For this reason, practically all primary magnets use superconducting coils.

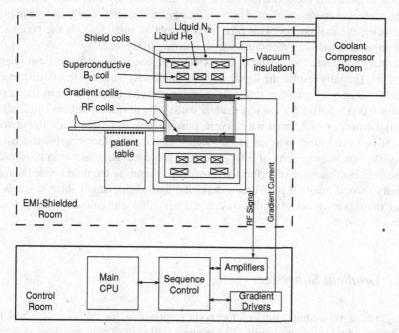

Fig. 5.22 Overview of a complete MRI system. EMF-critical components (the primary magnet, gradient-, and RF subsystems, and, of course, the patient) are placed in a separate, shielded room. Control units are found in a different room, from which the MRI scanner is operated. The cooling system that provides liquid nitrogen and liquid helium is placed in a third, separate room

A typical superconducting coil is a copper wire of about 2 mm diameter, which contains several NbTi filaments (strands) of about 0.1 mm diameter. This wire is then used to form an air coil (i.e., there is no supporting iron), and a typical MRI magnet coil can have a total filament length of several kilometers and carry currents of up to 500 A. The coil is then supercooled with liquid helium. One interesting aspect of the superconductive nature of the coil is the need to *charge* the coil, that is, introduce the current that then continues to flow without resistance. With a constant voltage U applied to the coil, the current builds up over time,

$$I_{Coil} = \frac{1}{L} \int U(t) \, dt = \frac{U}{L} \cdot t \qquad (5.37)$$

Once the target current is reached, the applied voltage is disconnected and the coil is shorted inside the supercooled system, thus allowing the current to continue to flow. A clinical 1.5 T magnet now has a total energy of 4 MWs stored, enough to power a 60 W light bulb for 18.5 h. When a magnet needs to be de-energized, it gets discharged just the opposite way, i.e., by applying a negative voltage until the current reaches zero.

The helium acts not only as supercoolant, but also as a safety medium in case of coolant failure. If the coil temperature rises above the superconductive threshold temperature, the energy of about 4 MWs is dissipated instantaneously, but the helium absorbs the thermal energy. This process is known as *quenching*. In the process, a large amount of helium evaporates, which is costly to replace.

A final important aspect of the B_0 coil is its homogeneity. Since magnetic gradients are used to spatially resolve the signal, any inhomogeneity leads to a spatial distortion. In fact, it is easy to calculate that a field inhomogeneity of only 50 ppm (parts per million, 0.005 %) in a 1.5 T clinical scanner with gradient system of 10 mT/m leads to a spatial distortion of 7.5 mm, which corresponds to tens of voxels or a shift between one and two axial slices in a typical 3D volume. Due to manufacturing tolerances, an out-of-the-box homogeneity of 500 ppm can be achieved. Any further improvement of the field homogeneity is performed on-site by measuring the field inside the bore (usually several hundred measurement points) and adding small shim magnets to offset the inhomogeneity. Additional shim coils allow extreme fine-tuning under software control.

5.7.2 Gradient Subsystem

The gradient subsystem contains the coils responsible for the G_x, G_y, and G_z gradients and their driver circuits. The gradient coil system contains three independent coils with independent drivers. Interestingly, off-axis gradients can be applied when two coils are driven simultaneously. This allows, for example, to acquire angled slices.

The main challenge for the gradient subsystem is the fast response time required for fast gradient switching (for example, the rapid on-off sequence for the phase encode gradient). To achieve large gradients and fast gradient rates at the same time, both the coil's inductivity and the resistance are kept low. Gradient coils require currents of several hundred amperes with a transient response of hundreds of kA/s. Accordingly, extremely powerful gradient drivers deliver high-voltage pulses to build up the current rapidly, and coil-amplifier systems are carefully compensated to provide the optimal step response. The compensation needs to take into account the generation of eddy currents in conductive structures around the coils.

The rapid switching of gradient fields inside a strong B_0 field causes the gradient coils to physically deform under the magnetic forces. This deformation causes the typical loud clicking, buzzing, or humming sound during the image acquisition.

5.7.3 RF Subsystem

The RF subsystem uses frequencies in the mid- to high MHz range, near the frequency range of FM radios. The resonance frequency for 1.5 T is 63.9 MHz, and for 3 T, 127.7 MHz, to provide two examples. Their corresponding wavelengths are 4.7 m and 2.35 m, respectively. In its simplest form, an RF coil is a cylindrical copper band with a small gap, like the letter C. The RF current, applied to the ends of the gap, causes a wave of current to travel along the coil. The resulting B_1 magnetic field rotates with the wave. The same coil can be used to receive the echo signal, but the sending and receiving currents are several orders of magnitude different. For higher sensitivity applications, transmit and receive coils are often separated, and the smaller receive coil is placed more closely to the organ under examination (e.g., head coil or knee coil). Coils are combined with a capacitor to create a resonant LC circuit.

More sophisticated coils include phased-array coils with multiple wires or multiple coils in a circular arrangement, which receive phase-shifted versions of the same signal. The result is again a rotating magnetic field. Three representative design principles are shown in Fig. 5.23. A single-wire coil with a capacitor to provide LC resonance (Fig. 5.23a) can be used both for transmitting and receiving RF signals. A similar coil, made of thin wire with multiple windings, is frequently used as a receive coil for surface applications. A resonator can provide a full quadrature signal. In Fig. 5.23b, six discrete antenna wires in the z direction carry sinusoidal currents with a 60° phase shift with the resulting current density rotating with the angle φ. Such a resonator is used for RF excitation only. The saddle coil (Fig. 5.23c) can be used for transmit and receive applications with small dimensions. The principle of the saddle coil is also used to generate the G_x and G_y gradient fields.

The RF signal is generated by digital synthesis. However, an enormous precision is required for the frequency synthesis. With single-side-band (SSB) modulation, a low-frequency offset can be added to a carrier signal at the Larmor frequency ω_0. An SSB modulator is a multiplier that takes the real and imaginary parts of the input signal $f(t)$ and multiplies it with the carrier frequency to obtain the transmit current

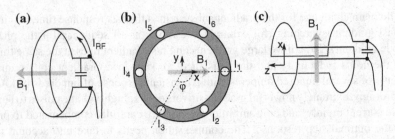

Fig. 5.23 Schematic of three possible realizations of an RF coil. **a** Single-turn solenoid with resonance capacitor. **b** Resonator cylinder (*axial view*) with discrete antenna wires. Each wire carries a sinusoidal current I_1 through I_6, each of which lags 60° over the previous one. **c** Saddle coil, typically used for small-sized receive coils

I_{RF}:

$$I_{RF} = \Re\{f(t)\} \cdot \cos\omega_0 t - \Im\{f(t)\} \cdot \sin\omega_0 t. \tag{5.38}$$

If the input signal is a low-frequency complex signal $f(t) = e^{j(\omega t + \varphi)}$, the resulting current is real-valued and contains both the low-frequency offset and its phase shift:

$$I_{RF} = \cos[(\omega_0 + \omega)t + \varphi]. \tag{5.39}$$

A transmit amplifier is required to amplify the current such that the resulting B_1 field is strong enough to allow short flip pulses. For example, $B_1 = 11.8\,\mu$T is needed to achieve a 180° flip in 1 ms (Eq. 5.11). Depending on the coil design, the peak output power of the transmit amplifier may exceed 2 kW.

The receive signal is quadrature-demodulated, meaning, the real-valued induction current is multiplied with $\cos(\omega_0 t)$ to obtain the real component of the signal and with $\sin(-\omega_0 t)$ to obtain the imaginary component of the signal. The multiplication provides terms with the sum and the difference of the frequencies. A low-pass filter removes components with $2\omega_0$, and the remaining signal contains only the low-frequency component (the component modulated by the frequency encode gradient!) and the phase. It becomes more and more common to use fast digital signal processors for demodulation, but the initial amplifier and low-pass filter need to be carefully designed to minimize amplifier noise.

Minimization of noise is also one reason why the entire MRI scanner, notably with its RF components, is housed in an electromagnetically shielded room (Fig. 5.22) to reject environmental interference from 60 Hz power lines, radio stations, cell phone towers, or other sources of electromagnetic contamination.

Chapter 6
Ultrasound Imaging

Abstract Ultrasound imaging makes use of the properties of sound waves in tissue. Pressure waves in the low megahertz range travel through tissue at the speed of sound, being refracted and partially reflected at interfaces. Ultrasound contrast is therefore related to echogenic inhomogeneities in tissue. The depth of an echogenic object can be determined by the travel time of the echo. By emitting focused sound waves in different directions, two-dimensional scans are possible. Ultrasound images are highly qualitative in nature due to the complex relationship between inhomogeneous tissue and the echoes, due to the differences in speed of sound in different tissues, and due to the high noise component that is a result of the weak signal and high amplification. Ultrasound images show good soft tissue contrast, but fail in the presence of bone and air. Although ultrasound images can be generated with purely analog circuitry, modern ultrasound devices use computerized image processing for image formation, enhancement, and visualization. Ultrasound imaging is very popular because of its low-cost instrumentation and easy application. However, an ultrasound exam requires the presence of an experienced operator to adjust various parameters for optimum contrast, and ultrasound images usually require an experienced radiologist to interpret the image.

6.1 Sound Propagation in Biological Tissue

Historically, ultrasound imaging emerged shortly after World War II from sonar (sound navigation and ranging) with somewhat similar technology. Sonar makes use of the excellent sound propagation properties of water, whereby a short acoustic pulse is introduced into the water. Any object in the path of the sound pulse causes a reflected echo, which can be picked up. Sound travels in water at approximately 1500 m/s, and the round-trip time of the echo gives an indication of the distance of the echogenic object. The first documented medical application was in 1942 [31], but major improvements of the instrumentation are likely attributable to Donald and

M. A. Haidekker, *Medical Imaging Technology*, SpringerBriefs in Physics,
DOI: 10.1007/978-1-4614-7073-1_6, © The Author(s) 2013

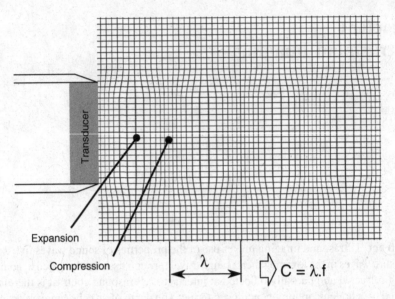

Fig. 6.1 Propagation of a compressive sound wave in elastic tissue. The sound wave is introduced by a piezoelectric element (transducer), and the wave propagates inside the tissue (*wide arrow*) with the speed of sound c, which is a tissue constant. The wavelength λ depends on the frequency f as $\lambda = c/f$

MacVicar in the 1950s [32, 33]. At this time, ultrasound imaging was purely based on analog electronics, and the image was visualized on a CRT screen. As such, ultrasound imaging is the only other imaging modality (apart from x-ray imaging) that does not *require* computerized data processing for image formation.

Like sonar, ultrasound imaging is based on the introduction of a compressive wave into the tissue (Fig. 6.1). Even the earliest instruments made use of the piezoelectric effect to generate the compressive wave. Wave propagation can be described by the partial differential equation

$$\frac{\partial^2 \zeta}{\partial t^2} = c^2 \cdot \frac{\partial^2 \zeta}{\partial z^2} \tag{6.1}$$

where $\zeta = (\rho - \rho_0)/\rho_0$ is the relative deviation of the tissue density ρ from its normal (uncompressed) density ρ_0, z is the direction of sound propagation, and c is the speed of sound. The one-dimensional nature of the sound wave is an approximation. Tissue viscoelasticity and diffraction broaden the sound beam, but the one-dimensional treatment is sufficient in the context of this chapter to examine the sources of contrast and methods of image formation. The solution of the one-dimensional wave equation is the complex harmonic oscillation

$$\zeta(z, t) = \zeta_{\max} \cdot \exp 2\pi j \left(\frac{z}{\lambda} - ft \right) \tag{6.2}$$

where λ is the wavelength (Fig. 6.1) and f is the frequency of the sound wave. A compressive wave implies the presence of a corresponding pressure, and the pressure depends on the amount of compression ζ and the *bulk modulus B* through $p = B \cdot \zeta$. The bulk modulus is inversely proportional to the compressibility of a material and can be seen as the material stiffness or resistance against deformation. For low ultrasound energies, linear elasticity is assumed (i.e., Hooke's law).

Two important tissue constants are the *speed of sound c* and the *acoustic impedance Z*, which depend on B and ρ_0 through

$$c = \sqrt{\frac{B}{\rho_0}} \; ; \quad Z = \sqrt{B \cdot \rho_0} \tag{6.3}$$

The acoustic impedance can best be explained as the ability of the tissue to conduct sound. In analogy to a voltage, which causes a current to flow, the sound pressure causes a local motion of the infinitesimal tissue elements with a velocity v. The velocity of a local particle is not the same as the speed of sound, but the two are related through $v = c \cdot \zeta$. To continue the analogy, Ohm's law relates voltage (pressure) to current (local velocity), and we find that $v = p/Z$. The same analogy leads to the definition of the *power* of the sound field as $P = v \cdot p$. Power is dissipated as the sound wave travels through tissue, and the power decreases from the incident power P_0 with traveled distance z according to Lambert-Beer's law,

$$P(z) = P_0 \, e^{-\mu z} \tag{6.4}$$

where the absorption coefficient μ depends on the tissue and the ultrasound frequency. It is practical to express sound attenuation A in decibels and rewrite Eq. 6.4 as

$$A = -20\text{dB} \, \log_{10} \frac{P(z)}{P_0} \approx 46 \cdot \mu_0 \, f \, z \tag{6.5}$$

where μ_0 is the attenuation coefficient at 1 MHz, f is the frequency, and z the distance traveled. As a rough approximation, sound power is decreased by 1 dB per cm and MHz in soft tissue. Some representative values of the attenuation coefficient, speed of sound, and the acoustic impedance are found in Table 6.1. The sound power lost by tissue attenuation is converted into heat, and at very higher power levels causes direct shear damage. Diagnostic ultrasound typically introduces less than 0.1 W/cm^2, and this level is thought to cause no harmful bioeffects. At higher power levels, the total exposure time would be limited to keep harmful effects to a minimum.

Ultrasound waves, like light waves, undergo reflection and refraction at interfaces with different acoustic impedance. At a smooth surface, the sound wave is split into a reflected and a refracted component that obey the law of reflection and Snell's law, respectively (Fig. 6.2):

$$\frac{\sin \theta_1}{\sin \theta_2} = \frac{c_1}{c_2} \tag{6.6}$$

Table 6.1 Some representative material and tissue constants for sound wave propagation

Material	Speed of sound c (m/s)	Acoustic impedance Z (kg/m^2 s)	Sound attenuation A (dB/MHz cm)	Half-value layer at 5 MHz (cm)
Air	330	430	–	–
Water	1492	1.49×10^6	≈ 0	–
Adipose tissue	1470	1.42×10^6	0.5	2.4
Liver	1540	1.66×10^6	0.7	1.7
Muscle	1568	1.63×10^6	2.0	0.6
Brain tissue	1530	1.56×10^6	1.0	1.2
Compact bone	3600	6.12×10^6	10 or more	0.12
PZT	4000	30×10^6		

PZT is a commonly used transducer material and is discussed in the next section

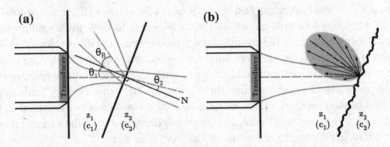

Fig. 6.2 Ultrasound propagation in inhomogeneous materials. **a** At a smooth interface, the sound wave is split into a reflected and a refracted component. The angle of incidence θ_1 with respect to the surface normal N is the same as the angle of reflection θ_R. The angle θ_2 of the refracted wave obeys Snell's law. **b** An irregular interface causes the reflected part to spread out in a diffuse pattern

The acoustic power P of the incident wave is split between the reflected and refracted part, and expressions for the reflectivity R and transmittivity T can be derived:

$$R = \frac{P_{\text{Reflected}}}{P_{\text{Incident}}} = \left(\frac{Z_2 \cos \theta_1 - Z_1 \cos \theta_2}{Z_2 \cos \theta_1 + Z_1 \cos \theta_2} \right)^2$$

$$T = \frac{P_{\text{Transmitted}}}{P_{\text{Incident}}} = \frac{4 Z_1 Z_2 \cos^2 \theta_1}{(Z_2 \cos \theta_1 + Z_1 \cos \theta_2)^2} \tag{6.7}$$

In biological tissue, the occurrence of an irregular surface is more likely that causes diffuse reflection (Fig. 6.2). In addition, inhomogeneities that are smaller than the wavelength of the sound wave cause Rayleigh scattering. Reflected or scattered sound waves that return to the transducer can be recorded as an echo and used for image formation. However, the energy loss through wave reflection is even more significant than energy loss through regular tissue attenuation, and the echo signal can be very weak. To get an idea of the amount of energy loss, let us assume an incident sound wave normal to a muscle-fat interface. By using the values from Table 6.1 and the reflectivity in Eq. 6.7, we can calculate that only about 0.1 % of the incident power is reflected back at the transducer.

Another similarity between light waves and sound waves is the beam expansion by diffraction. Without covering the theory of sound propagation in three dimensions in detail, the beam expands approximately linear with the traveled distance. The beam diameter d can be described by the *Fraunhofer approximation*,

$$d(z) = \frac{\lambda}{D} \cdot z \qquad (6.8)$$

where D is the diameter of the transducer, which is assumed to be circular. The region where the Fraunhofer approximation holds begins at a distance of approximately D^2/λ from the transducer and is called the *far-field* region. At the opposite end, when $z < D^2/\lambda$, the sound pressure pattern becomes very complex. The beam diameter converges and forms a waist, and interference causes the formation of side lobes. This region is called the *near-field* and is generally avoided for ultrasound echo generation. For a general approximation, it is sufficient to assume a constant near-field beam diameter $d = D$.

6.2 Ultrasound Image Formation

6.2.1 Ultrasound Generation and Echo Detection

The key element in ultrasound imaging is the sound transducer. The transducer is made of a piezoelectric material. Most frequently, lead zirconate titanate (PZT) is used, which is a ceramic combination of $PbZrO_3$ and $PbTiO_3$ molecules. Piezoelectric materials consist of strong dipoles that change the shape of the crystalline structure when exposed to an electrostatic field. Conversely, when the crystalline structure is exposed to mechanical stress, the crystal exhibits an electrostatic potential across its sides [34, 35]. Therefore, the transducer can serve both as pulse generator and microphone for the sound echoes. Typical ultrasound scanners use several hundred volts to create a surface displacement in the μm range. A retuning echo causes a potential in the nanovolt to microvolt range.

Transducers are operated in resonance, with part of the wave energy reflected back into the PZT material. A standing wave forms at the fundamental resonant frequency f_0,

$$f_0 = \frac{c_{PZT}}{\lambda_0} = \frac{c_{PZT}}{d_{PZT}} \qquad (6.9)$$

where c_{PZT} is the speed of sound in PZT and d_{PZT} is the thickness of the PZT layer. The resonance frequency is therefore determined by the thickness of the PZT transducer element. A 5 MHz transducer, for example, can be obtained with a 0.8 mm PZT

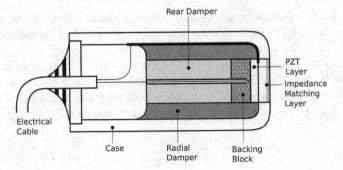

Fig. 6.3 Sketch of a simple single-element ultrasound transducer. A thin layer of PZT is the active element. Toward the tissue, an impedance-matching layer is responsible for minimizing the loss between the high-impedance PZT and the relatively low-impedance tissue. A backing material ensures that the wave is emitted through the acoustical window, and the damper attenuates the resonant vibrations and therefore controls the pulse duration

layer. The resonant oscillation is excited with a short voltage pulse of microsecond duration.

A sketch of a single-element ultrasound transducer is shown in Fig. 6.3. The PZT layer, together with the backing material and the impedance-matching layer, forms the actual transducer element. Absorber material (i.e., the acoustic damper) in the ultrasound probe determines the attenuation of the resonant oscillation and therefore the duration of the ultrasound pulse. Furthermore, the attenuation determines the bandwidth of the pulse. Strong attenuation causes a broader bandwidth than weaker attenuation, and consequently a shorter pulse. A broad bandwidth (short pulse) increases the axial resolution at the expense of SNR. A practical example of a transducer element is shown in Fig. 6.4.

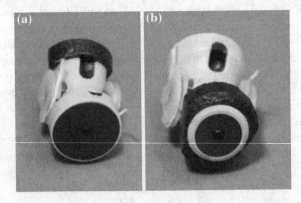

Fig. 6.4 Photo of a single-point dual-focus transducer element. Shown is the front view (**a**) and back view (**b**) of a movable element that carries two transducers. Both transducers are slightly curved to produce a converging sound wave, and the smaller transducer has a higher curvature and therefore a focal point nearer to the transducer

Table 6.2 Ultrasound frequency, depth and resolution limits, and typical clinical applications

Ultrasound frequency (MHz)	Maximum depth (mm)	Axial resolution Δz (mm)	Lateral resolution Δx (mm)	Typical application
3	150	0.6	2.0	General purpose; fetus, heart, liver
5	100	0.35	1.2	Kidney, heart, brain
10	50	0.2	0.6	Muscle, tendons, endoscopic applications (prostate)
15	33	0.15	0.4	Intraoperative applications blood vessels
≥ 20	≤ 25	≤ 0.1	≤ 0.3	Research applications; vasculature, skin

At its acoustical window, an impedance-matching layer is placed between the PZT crystal and the tissue. For single-layer designs, the impedance-matching layer is a polymer with the approximate acoustic impedance $Z = \sqrt{Z_1 \cdot Z_2}$, where Z_1 is the acoustic impedance of PZT and Z_2 that of tissue. The additional application of ultrasound gel is necessary to remove any air between the transducer and the tissue. The acoustic impedance of air is extremely low (Table 6.1), and more than 99.9 % of the sound wave would be reflected at any tissue-air interface.

The resonance frequency of the transducer depends on the clinical application. The wavelength of the sound wave provides the theoretical limits for the axial resolution ($\Delta z \geq \lambda/2$) and also for the lateral resolution through the numerical aperture, where under ideal conditions (NA $= f_0/D \approx 1$) the lateral resolution $\Delta X \geq \lambda$. Higher frequencies therefore provide better spatial resolution, but the attenuation coefficient also increases, and maximum depth is reduced. Table 6.2 lists commonly used ultrasound frequencies with their approximate depth and resolution limits, and some applications.

6.2.2 A-Mode Scans

The simplest form of an ultrasound scan is to emit a short sound pulse and wait for the arrival of an echo. The round-trip travel time is converted into the depth z of the echogenic object, and the echo amplitude is drawn in a two-dimensional coordinate system over the depth z. This scan is referred to as *A-mode scan* or amplitude-mode scan. The strongest echoes are generated along the beam path, and the beam diameter determines the lateral resolution. A schematic representation of an apparatus to generate a sound burst and record the echoes is shown in Fig. 6.5. Typically, the same transducer is used for sound wave generation and for recording the reflected waves. A master timing generator is responsible for switching the transducer between

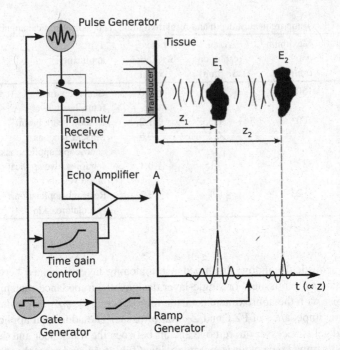

Fig. 6.5 Schematic representation of an instrument to generate an A-mode scan in pulse echo operation. A master timing generator (gate generator) repeatedly switches the transducer to transmit mode and applies a pulse waveform through a pulse generator, thus causing a short sound pulse to travel into the tissue. The transducer is immediately put into receive mode, and echoes are recoded and plotted along the time axis. By multiplying the time axis with $c/2$, it can be scaled to provide the approximate depth z. A ramp generator provides a signal that is proportional to z. The echo amplitude A is corrected for sound attenuation by increasing the gain exponentially with time (time gain control). In this example, two echogenic objects E_1 and E_2 are sketched at depths z_1 and z_2, respectively, and their echoes are recorded as the amplitude A over the time axis

transmit and receive modes, for causing a signal generator to generate the sound signal, and for starting the time-resolved echo acquisition. This mode of operation is also referred to as *pulse echo acquisition*, because the transmitter is used only for short periods of time to emit a sound pulse. The round-trip time t_{rt} of an echo is directly related to its depth z through $z = c \cdot t_{rt}/2$. The ultrasound scanner cannot know what type of tissue is traveled by the sound wave, and some representative value of c, usually 1540 m/s, is assumed when quantitative measurements are made.

The master timing generator operates with a *pulse repetition frequency* of typically 2–4 kHz, and the pulse repetition frequency determines the maximum depth of the scan. For an average speed of sound $c = 1540$ m/s, a 4 kHz pulse repetition frequency translates into a maximum round-trip time of 250 μs and a maximum depth of 19 cm. With increasing depth, the sound wave is attenuated. The receive amplifier therefore increases its gain exponentially with increasing time, and the gain is reset upon the next pulse. This correction is called *time gain control*. If we use the rule-of-thumb attenuation of 1 dB per cm and MHz and a 5 MHz probe, the time gain control unit would add 10 dB gain at 2 cm depth (26 μs round-trip time), and 50 dB at 10 cm (130 μs round-trip time).

In multidimensional scans (B-mode and M-mode scans), the pulse repetition frequency determines the total image acquisition time. However, very high pulse repetition frequencies can lead to aliasing when an echo of a strongly echogenic object arrives later than the next pulse is emitted. For example, the echo from a reflecting interface at 12 cm depth arrives after approximately 156 μs. The pulse repetition time for an 8 kHz pulse repetition frequency is 125 μs, and the echo would arrive 31 μs after the subsequent pulse. The ultrasound scanner would draw this echo as a weak pulse from an object approximately 2.4 cm deep.

6.2.3 B-Mode Scans

The amplitude of an A-mode scan can be encoded as intensity (brightness) in one line of an image. Lateral movement of the transducer would then provide the second dimension, and a two-dimensional image is presented. This type of scan is termed *B-mode scan* for its brightness encoding. In many scanner designs, the transducer movement is indeed mechanical, as shown in Fig. 6.6. Two dominant designs exist, the first where the head performs a wobbling motion (Fig. 6.6), and the second, where multiple transducers are arranged on the outer surface of a rotating cylinder.

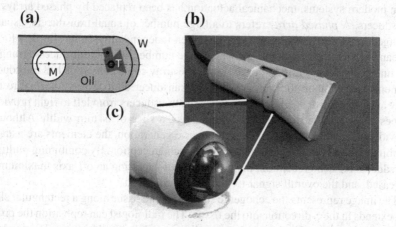

Fig. 6.6 Focused ultrasound source mounted on a mechanical pivot to provide a B-mode scan. **a** Sketch of the scan head. The transducer T emits a focused beam through the acoustical window W. The transducer assembly is mounted on a pivot and driven by a motor M, thus allowing to position the beam along a fan-shaped plane. The entire assembly is immersed in index-matching oil that allows ultrasound transmission through the acoustic window. **b** Image of a scan head that is built after this principle. From the acoustic window, the sound is guided along a soft silicone pad with the surface that actually comes in contact with the tissue. If the silicone pad is removed (**c**), the oil-immersed transducer and its mounting mechanism become visible. The transducer element itself is similar to the one shown in Fig. 6.4

(a)   **(b)**

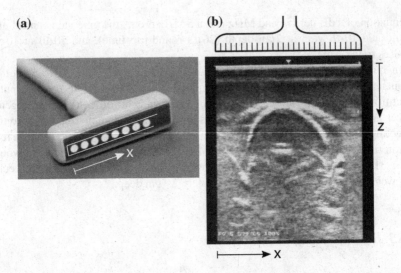

Fig. 6.7 Linear transducer array for high-resolution ultrasound imaging. **a** The matching layer covers a large number of individually controllable transducer elements (indicated as white circles, which are not to scale). Each element sends out an ultrasound wave with its respective lateral position x. By pulsing the elements in sequence, the individual A-mode scans can be arranged into a two-dimensional image. **b** The B-mode image shows a cross-section through a cotton boll immersed in water

In modern systems, mechanical actuation has been replaced by phased arrays of transducers. A *phased array* refers to a large number of small transducers, usually arranged in a linear fashion, that can be driven independently. In its simplest form, we can envision the transducer array as a large number of spot transducers arranged in a line as shown in Fig. 6.7. For example, the array could consist of 128 transducer elements spaced about 0.5 mm apart. Each transducer emits a wave normal to the array at its respective position, and pulsing the transducers from left to right provides echoes along parallel lines (Fig. 6.7b) to cover a slice of 64 mm width. Although individual elements are pulsed for sound wave generation, the elements are usually combined (that is, their signals added up) for echo reception. By combining multiple transducers for echo reception, the probability of capturing an off-axis maximum is increased, and the overall signal-to-noise ratio improved.

The image represents the echogenic strength of the tissue along a rectangular slice that extends in the z-direction into the tissue. The radiologist can reposition the probe or change the angle of incidence to cover different planes in the tissue. The need for continuous manual interaction is one of the most fundamental differences between ultrasound imaging and other imaging modalities.

The example of the linear array is one possible application of phased arrays. Another important aspect is the ability of phased arrays to shape and direct the sound beam. The idea is to generate a curved sound front. The transducer elements are closely spaced, and are considered to generate a spherical wave. For example, a larger curved transducer (such as the point transducer in Fig. 6.4) can be replaced by a phased array where the outer transducers are pulsed earlier than the inner ones (Fig. 6.8a). More precisely, a spherical wave front can be obtained with a delay $\tau(x)$ between the beginning of the pulse at $t = 0$ and the pulsing of an element at position x,

$$\tau(x) \approx \frac{1}{c} \cdot \frac{(l/2)^2 - x^2}{z_F} \qquad (6.10)$$

where l is the overall length of the linear array and z_F is the position of the focus along the z-axis. With the appropriate asymmetrical time delays, the center of the spherical sound front can be shifted away from the z-axis, thus allowing the resulting sound front to enter the tissue at an angle θ with respect to the z-axis (Fig. 6.8b). This angle can be changed between each A-mode scan, and the resulting echoes are collected from a wedge-shaped plane. In a similar fashion, a *receive focus* can be achieved by applying a short position-dependent delay to the echo signals of each transducer element before the summation of their signals.

6.2.4 M-Mode Scans

The M-mode scan (M stands for *motion*) is achieved by arranging intensity-encoded A-mode lines as a function of time. Each line of the M-mode scan therefore represents the depth of an echogenic object along the z-axis, and its motion along the z-axis can be imaged. M-mode scans are often obtained to observe the motion of the heart valves, and the evolution of the M-mode scan along the time axis can be correlated to the ECG of the patient.

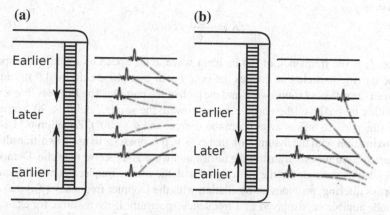

(a) **(b)**

Earlier

Later

Earlier

Earlier

Later

Earlier

Fig. 6.8 Illustration of the beamforming capabilities of a phased array. **a** Focus generation. By pulsing the outer transducer elements earlier than the center elements, a curved sound front is achieved that converges towards a focus. The location of the focus can be determined by the phase difference between the pulsed elements. **b** Beam steering. Asymmetric pulsing of the transducer elements leads to a sound front that is not symmetrical and therefore travels at an angle with respect to the z axis

6.2.5   Volumetric Scans and 3D Ultrasound

Ultrasound scans can be extended into three dimensions by mechanical actuation of the transducer along a third axis or with two-dimensional phased arrays that allow beam steering in two dimensions. This technique has become very popular for diagnosing fetal congenital diseases [36, 37]. The amniotic fluid has a very low echogenicity and is followed by a strong echo from the fetus. Image processing and surface reconstruction allow a very photo-like rendering of the fetus.

Often, reference is made to 4D ultrasound [38] . The fourth dimension is time, and 4D ultrasound imaging refers to an acquisition process fast enough to capture motion. When we consider a typical 4 kHz pulse repetition rate, a B-mode scan with 512 A-mode lines has a frame rate of $8s^{-1}$. If a third dimension is added, for example, with 128 B-mode scans, the 3D image acquisition time is approximately 16 s. This time is too long to allow the usual manual operation of the scanner. Initially, image acquisition time was reduced by reducing the number of pulsed elements [39], with consequently reduced resolution and image quality. With narrower angular coverage and faster transducers, significant speed gains can be realized without sacrificing image quality [40]. New approaches, such as single-wave coherent imaging and shear wave imaging promise even faster imaging speeds [41].

6.3   Doppler Ultrasound

One of the most powerful features of ultrasound imaging is the ability to directly measure blood flow by exploiting the Doppler effect. The frequency of a sound wave reflected by a moving object differs from the incident frequency by the *Doppler frequency* f_D,

$$f_D = \frac{2 f_0 v}{c} \cdot \cos \theta \qquad (6.11)$$

where f_0 is the frequency of the incident wave, c the speed of sound, v the speed of the moving particle that creates the echo (such as a blood cell), and θ the angle between the reflected sound wave and the path of the particle. The Doppler frequency can either be positive (the particle moves toward the source, and $\theta < 90°$) or negative (the particle moves away from the source, and $\theta > 90°$). Equation 6.11 is an approximation with the assumption that $c \gg v$. It is possible to use a two-transducer system to continuously transmit the ultrasound wave and receive the echo. Demodulation, that is, multiplying the echo signal with the carrier frequency and subsequent low-pass filtering, provides an oscillation with the Doppler frequency $|f_D|$ that can be made audible or displayed as a frequency spectrum. If we assume, for example, a 3.5 MHz probe and blood velocity of 15 cm/s with the probe at a 45°angle to the direction of blood flow, the Doppler signal would have a frequency of roughly 500 Hz (Eq. 6.11). This frequency is well within the audible range.

More commonly used is the pulsed Doppler operation, because the Doppler velocity measurement can be combined with the ability of a regular pulsed-operation
ultrasound system to provide a spatially resolved signal. Devices operating in pulsed
Doppler mode exhibit three differences from regular B-mode scanners:

- The acoustic damper in the transducer element is usually weaker, creating a more
  sustained wave. With the longer transmit pulse, some axial resolution is sacrificed,
  but a better signal detection is possible
- The maximum A-mode depth is reduced and the pulse repetition frequency raised.
  The reason for choosing a higher pulse repetition frequency is discussed below.
- The signal used for Doppler detection is gated, that is, only signals arriving during a
  short time window between two subsequent pulses are used for Doppler detection.
  The time window corresponds to a depth section and is selectable by the operator.

In pulsed Doppler operation, the phase shift between incident wave and echo is
determined. The phase shift depends both on the travel time of the echo (i.e., the
depth of the object) and the velocity. With short pulse repetition times, the object
can be assumed to be stationary. A small blood volume traveling at 15 cm/s, for
example, moves only 37 μm between two pulses at a pulse repetition frequency of
4 kHz. This is a fraction of the lateral resolution. If the flow velocity is assumed to
be constant in the brief period between pulse and echo, the Doppler frequency f_D
can be recovered from the measured phase shift φ through $f_D = \varphi/(2\pi t) - f_0$.
With suitable quadrature demodulators, the sign of φ can be recovered, allowing to
determine the flow direction, and the quadrature signal can be analyzed by computing
the Fourier spectrum or by correlating signals of two subsequent pulse echoes. Since
the echo phase is sampled only once per pulse repetition period, the resulting velocity
signal is time-discrete, and the maximum Doppler frequency that can be recovered
is exactly one half of the pulse repetition frequency (Nyquist limit). By using this
relationship between f_D and the pulse repetition frequency f_p, the maximum flow
velocity can be computed with Eq. 6.11:

$$v_{\max} = \frac{c \cdot f_p}{4 \cdot f_0 \cdot \cos\theta} \tag{6.12}$$

Any higher flow velocity is subject to *aliasing*, that is, the phase detector
incorrectly indicates a much lower velocity. A higher pulse repetition frequency (and
consequently a lower scan depth) allows to measure proportionally higher velocities.
The incident angle is also critical in Doppler ultrasound measurements. For example,
a probe held at a 45°angle can cause an error of ±6 % when the probe angle is var-
ied as little as ±3°. The error rises disproportionately with higher incidence angles.
Those are undesirable, though, because the Doppler frequency is lowered. However,
sometimes the radiologist needs to find a compromise between a high incidence angle
that is needed to access certain blood vessels and measurement accuracy.

Since the results of the Doppler measurement are spatially resolved, flow velocity
can be superimposed over the B-mode scan. This mode is sometimes termed *duplex
ultrasound scanning*. Velocity is color coded and superimposed over the grayscale

B-mode scan. By convention, flow in the direction of the scan head is colored in
red shades, and flow in the opposite direction in blue shades. Pathological flow, for
example a reversed flow direction in a bifurcation or stenosis, becomes immediately
evident, even if the absolute flow velocity is subject to measurement error.

Chapter 7
Trends in Medical Imaging Technology

Abstract Medical imaging technologies have, to a varying extent, experienced significant recent progress. From the introduction of a new imaging modality to its adoption to routine clinical practice, many years of development and testing are needed. Recently, the focus of medical imaging research shifted toward detail optimization and the development of new disease-specific protocols, although several striking new developments need to be highlighted. For example, the adaptation of phase and darkfield contrast, well-known from microscope imaging, to X-ray imaging, provides a new and astonishing level of contrast in X-ray imaging and CT. Another example is the development of new, ultra-portable ultrasound devices, which make ultrasound even more attractive as a fast and low-cost imaging modality. Laser-based optical imaging that uses visible or near-infrared light deserves special attention as some methods have been adopted in clinical practice. Last but not least, improved image processing—both in terms of new algorithms and of improved computing power—have continually improved image quality and opened new avenues of image processing, with many new functions available to aid the radiologist in providing a diagnosis.

The enormous technological progress seen in the last decades of the twentieth century gave rise to what could be called pioneering days of tomographic modalities. Invention of the transistor, the integrated circuit, and eventually the microprocessor were prerequisites for the development of modern imaging methods. During the same period of time, many of the fundamental image processing algorithms were introduced, such as digital filters or automated segmentation. Medical imaging technology benefited from these developments, and in the course of about two decades, the image quality delivered by CT, MRI, and ultrasound devices increased in great strides.

In recent years, progress has become more detail-oriented, with major efforts dedicated to optimizing imaging protocols for specific organs or diseases. Two examples can serve to illustrate this trend:

- Spiral CT (introduced in 1990 [42]): Conventional CT scanners complete one revolution to acquire a slice before advancing to the next axial slice. Spiral

M. A. Haidekker, *Medical Imaging Technology*, SpringerBriefs in Physics,
DOI: 10.1007/978-1-4614-7073-1_7, © The Author(s) 2013

(or helical) CT differs in that the patient table is advanced during the rotation of the source-detector gantry. The reconstruction maintains the same slice by inter-polating between two successive rotations. The main advantage is that slices at arbitrary axial positions, even overlapping slices, can be reconstructed from the same raw data set. The development of the helical scanning principle was accom-panied with improved cooling of the X-ray tube and improved detector efficiency, as well as improved gantry mechanics (the slip-ring gantry), which overall allowed one revolution of the gantry to be completed in one second. With such fast acqui-sition rates, motion artifacts were reduced, and complete volumetric scans could be completed in one breath hold.

• Open MRI (introduced in 1992 [43]): The open MRI scanner uses a C-shaped magnet with the primary B_0 field following a vertical orientation. In comparison, conventional MR scanners use a toroidal coil with a horizontal field orientation. The main advantage of the open MRI geometry is easier access to the patient during imaging, which facilitates, for example, interventional imaging. Open MRI enjoys additional popularity, because it does not expose the patient to the somewhat claustrophobic environment of the conventional magnet bore.

In the following sections, some recent developments are highlighted. In short, the trend of improving existing modalities continues. Some of the progress is made by combining multiple imaging modalities, and by obtaining functional information with slightly altered imaging protocols. Optical imaging, that is, imaging with visible or near-infrared light, is a relatively young modality that is still in the development stages. In all cases, advances are being helped by progress in computerized image processing and image understanding.

7.1 Progress in Established Imaging Modalities

7.1.1 X-ray and CT

In recent years, X-ray imaging has reached a technological plateau. The trend away from film and toward digital X-ray imaging continues. Improved detectors with higher sensitivity allow to further reduce exposure time and the patient radiation dose. After concerns were raised that increased use of X-rays in diagnostic and interventional procedures could lead to elevated cancer risk (see, e.g., [44]), a shift away from X-ray imaging toward ultrasound and MRI has been observed, leading to a further reduction of the radiation exposure in patients. Computed tomography, however, remains an attractive modality because of its very high per-slice acquisition rates, notably with the development of dual-source CT scanners [45]. Modern dual-source CT scanners are capable of 0.3 s or less per rotation with an axial speed of 0.4 m/s. With such high acquisition speeds, motion artifacts cease to be a concern. In addition, the heart can be scanned in 3D during one heart beat. Lastly, modern CT scanners give rise to sub-mSv scans (i.e., scans with a total exposure of less than 1 mSv, which approaches the level of a typical chest X-ray of 0.1 mSv). Today, radiation exposure from CT scans is less of a concern than 20 years ago.

Transmission-based X-ray imaging has recently been complemented by phase-contrast and darkfield methods that are known from light microscopy. Phase contrast microscopy makes use of wave interference: The illumination light wave is split in two parts, a reference wave and a wave that experiences a phase delay in the object (in media of different refractive index, the apparent path length is changed). In a similar way, X-rays experience a phase change in weakly absorbing materials [46]. With suitable diffraction gratings, the phase (more precisely, its first derivative perpendicular to the grating slits) can be converted into intensity and thus recorded by the detector [47]. The same principle can be used to record scattered X-rays, leading to the X-ray analog of darkfield imaging [48].

Phase contrast X-ray imaging provides the projection of the refractive index along the beam path, analogous to conventional X-ray imaging that provides the projection of the absorber density. Therefore, phase contrast-enhanced radiography advertises itself for CT reconstruction methods [49, 50]. These methods promise not only markedly enhanced perception of contrast, but actually a different type of information retrieved from the scanned object, namely, its refractive index. Particularly in conjunction with CT reconstruction methods, tissue-tissue contrast could be dramatically enhanced, thus eliminating one weakness of X-ray based CT imaging. However, the method is still under development, and it will probably take several years before phase contrast CT can be found in medical diagnostic centers.

7.1.2 Magnetic Resonance Imaging

Magnetic resonance imaging experiences progress from the use of stronger magnets and improved amplifiers. Both lead to higher spatial resolution and improved SNR, or, with constant SNR, to shorter acquisition times [51]. A significant milestone was the introduction of a blood-oxygen level dependent (BOLD) sequence [52]. The BOLD sequence makes use of differences in T_2^* relaxation times between oxyhemoglobin and deoxyhemoglobin and allows to measure blood flow and blood oxygenation levels. This technique has given rise to *functional MRI*. Although functional MRI enjoys most of its popularity in studies to localize brain activity, clinical applications exist, including Alzheimer's disease and the measurement of coronary blood flow. Due to its low SNR and lower spatial resolution, BOLD functional images are often superimposed over structural MRI images, similar to PET and SPECT images.

Diffusion tensor imaging is another method to take MRI in the direction of functional imaging. The Bloch equation (Eq. 5.15) assumes stationary protons; an additional term that depends on diffusion modulates the change of magnetization dM/dt. Diffusion tensor imaging allows to recover the three-dimensional vector field of water diffusion [53]. Clinical applications include brain imaging in dementia, including Alzheimer's disease [54].

Another "abnormal" MR imaging technique makes use of reduced T_2^* as a consequence of microscopic susceptibility changes caused by bone [55]. Normally, MR is not widely popular for bone imaging due to the low proton density and the short T_1 and T_2 relaxation times in bone. Special sequences, such as FLASE (fast, low-angle spin echo), make use of T_2^* contrast in bones, while achieving voxel sizes between 0.1 and 0.2 mm [56]. Although bone strength assessment widely relies on ultrasound (for a fast, less accurate measurement) and X-ray imaging (for a higher-accuracy assessment), micro-MRI imaging of bone promises to evolve into one pillar of trabecular structure assessment [57]. These examples highlight the present trend in MRI to extract more information from the tissue with the same underlying physical principle, but with the sophisticated application of new sequences.

7.1.3 Ultrasound Imaging

Ultrasound imaging technology has also reached a certain plateau. Ultrasound imaging remains the modality of choice for rapid, low-cost diagnostic procedures without ionizing radiation. The size of ultrasound scanners has been reduced dramatically over the last two decades, and recently hand-held ultrasound scanners with full diagnostic capabilities were introduced [58]. It can be expected that the popularity of ultrasound imaging will further increase with the spread of ultra-portable devices.

Ultrasound contrast agents have been introduced that consist of gas-filled microbubbles. These bubbles literally burst in the incident sound field and create a strong signal. These microbubbles can be functionalized to bind to specific sites, such as tumors or inflammatory processes [59]. With such contrast agents, ultrasound, too, takes the step toward functional imaging, with the additional potential for targeted drug delivery [60] as the microbubbles can be loaded with drugs, set to burst at the insonicated target site.

7.1.4 PET and Multi-Modality Imaging

PET offers one potential improvement that SPECT cannot offer: more precise localization of a decay event with time-of-flight measurements, and a resulting improvement of SNR. Ultrafast electronics and fast-decay scintillation crystals are a prerequisite. For example, to achieve a time-of-flight resolution of 45 mm, the detector needs a resolution of 150 ps, which is technologically feasible [61]. Moreover, experiments with semiconductor detectors are under way. The proposed detector element is an avalanche photodiode, which has a much higher sensitivity than CMOS or CCD elements, but a lower sensitivity than a PMT. Avalanche photodiodes are much smaller than PMTs and promise dramatically improved spatial resolution.

Combined SPECT/CT devices and PET/CT devices have been introduced more than 10 years ago [62], a technology that advertises itself, because the detectors for gamma and X-ray radiation are similar. These scanners remove the need for image fusion related to moving a patient between different scanners. Technologically more challenging is the combination of PET with MRI, and the first PET/MRI multi-modality scanners became available only in the last few years [63]. Once again, the development of semiconductor detectors was crucial, because the strong magnetic field of the MRI device makes the use of PMTs impractical.

7.1.5 Molecular Imaging

In the context of radionuclide imaging and multimodality imaging, brief mention of molecular imaging is appropriate. Molecular imaging can be defined as imaging of functional (i.e., physiological) processes at the cellular and subcellular scale. Traditionally, fluorescent markers in conjunction with light microscopy were used for research at this level. The concept of molecular imaging can be extended to high-resolution functional imaging with tomographic methods. SPECT and PET can be used with radiolabeled antibodies or other markers that specifically interact with targeted proteins (see, for example [64] for a review of molecular imaging in cardiovascular applications). Another emerging field links molecular imaging with stem cells: Stem cells are loaded with paramagnetic nanoparticles, which allows them to be traced with MRI [65].

In Sect. 7.1.3, we briefly introduced functionalized microbubbles for functional ultrasound imaging. Another related imaging technique is photoacoustic imaging and photoacoustic tomography [66]. In the case of photoacoustic imaging, the sound wave is generated inside the tissue by absorption of a high-energy pulse of light. Absorption can take place either in intrinsic chromophores (e.g., hemoglobin) or in externally supplied dyes. B-mode scan techniques can be used for spatially resolving the sound source, or array detectors allow tomography-like reconstructions. One of the limits of photoacoustic imaging is light scattering, which limits the depth of the incident light pulse.

7.2 Optical Tomography

Optical tomographic imaging modalities are those that use visible or near-infrared light for image formation. They deserve a separate section, because optical imaging currently finds its way into medical practice. Optical imaging is attractive, because it does not use ionizing radiation, has short acquisition times, offers a spatial resolution much higher than ultrasound imaging, and potentially can be achieved with low-cost instrumentation.

Visible or near-infrared light experiences strong scattering in biological tissue, and common optical imaging modalities are limited to a depth of few millimeters. With the exception of optical transillumination tomography, optical imaging cannot rely on unscattered photons, and reconstruction methods such as those in CT cannot be used.

Optical coherence tomography (OCT) can be considered the optical equivalent of ultrasound. The light source is a special type of laser with an unusually broad bandwidth. Normally, lasers are considered to be monochromatic, that is, their emission spectrum is very narrow. Conversely, OCT lasers (known as superluminescent diodes, SLD) have a broad bandwidth, more precisely, a Gaussian wavelength distribution with a bandwidth σ of 20–100 nm. Broadband light loses its coherent properties rapidly, and a superluminescent diode has an approximate coherent length of $L_c \approx 0.44\lambda^2/\sigma$.

Optical coherence tomography instrumentation is based on the Michelson interferometer. The sketch of a basic OCT system shown in Fig. 7.1 is based on free-space optics, although OCT devices are normally based on fiber optics. In either case, the light from the SLD is split into a sample and a reference path. Light is scattered back from the tissue and recombined with the reflected reference beam. Only light scattered from the tissue section where the reference and sample beams have the same length are recorded by the photodetector. By moving the reference mirror, the depth of the detected light is changed. Moving the reference mirror therefore produces a scan of the scattered light amplitude $A(z)$, which is the optical equivalent of an ultrasound A-mode scan.

Due to the short coherent length of the SLD, the axial resolution is high, often in the range of 5–15 μm. Depending on the focusing optics (not shown in Fig. 7.1), similar resolution can be achieved in the lateral direction. A scan mirror in the sample beam path can deflect the beam and sweep it much like an ultrasound beam (cf. Fig. 6.6)

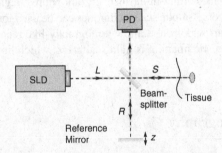

Fig. 7.1 Schematic of an optical coherence tomography device. The SLD emits a collimated beam of light with short coherent length. A beamsplitter (i.e., a semi-transparent mirror) splits the beam into the sample path S and the reference path R. Scattered light from the tissue can only be recorded at a depth where the reference arm and sample arm of the beam have the same length. Scattered light is sent back to the beamsplitter, where it is recombined with the reflected reference beam and directed onto a photodiode (PD). Moving the reference mirror (z) moves the position of the coherent signal and produces the optical equivalent of an A-mode scan

to produce a B-mode scan. The source of contrast is the amount of light scattered by the tissue back along the sample path.

Even more elegant is Fourier-domain OCT. It can be shown that the wavelength of the light (more precisely, its frequency) contains the Fourier-encoded depth information. In other words, frequency $\nu = c/\lambda$ and depth z are related through the Fourier transform. To make use of this principle, the reference arm in Fig. 7.1 is movable only for coarse depth adjustment, and the detector is replaced by a spectrometer. As a consequence, the scattered signal $s(\nu)$ is resolved by the frequency. Inverse Fourier transform of $s(\nu)$ yields the scattered intensity $A(z)$. The advantage of Fourier-domain OCT is its ability to obtain a complete A-mode scan in one measurement. Its main disadvantage is the lower SNR that is a consequence of the scattered light being distributed over many detectors in the spectrometer.

OCT has found its way into clinical practice, primarily used by ophthalmologists to examine the retina [67] and less frequently the cornea [68]. In dermatology, OCT scans help diagnose skin cancer and inflammatory processes [69]. The flexibility of fiber optics also allows OCT to be used in the gastrointestinal tract [70].

Diffuse optical tomography (DOT) uses scattered photons to reconstruct the scattering coefficient in the sample. A typical DOT imaging device uses a cylindrical sample holder filled with index-matching fluid. Along two closely-spaced circles are arrayed a ring of point sources (optical fibers connected to a laser) and point detectors (optical fibers connected to photodetectors). For a known medium and known geometry, the wave propagation equation can be solved, and the intensity at the detectors predicted. However, unlike tomographic methods that are based on straight-line geometries (CT), no closed-form solution for the inverse problem exists, and iterative methods need to be employed to reconstruct the geometry from measured light intensities [71]. The method can be modified to measure the local concentration of fluorescent emitters [72]. With fluorescently labeled drugs or physiologically active compounds, DOT holds the promise to image physiological processes analogous to SPECT and PET. The main disadvantage of the method is its poor spatial resolution. A commercial DOT-based screening device for breast cancer has been introduced recently, and DOT is in the process of being adapted for clinical practice [73, 74].

Optical transillumination tomography is the optical equivalent to X-ray CT. The single most difficult challenge for optical transillumination tomography is any form of refractive index change along the path, which invalidates the straight-line assumption of the Radon transform. In addition, the strong scattering properties of tissue require a very high dynamic range photodetector. Optical transillumination tomography has been proposed for several decades, but has not yet found applications in clinical practice [75]. Some attempts have been made to correct for refractive index mismatch [76], but those were limited to well-defined geometries, such as tissue-engineered blood vessels. However, in this context, optical transillumination tomography can offer unusually high acquisition speed [77]. A different approach is to use in vitro preparations that reduce the index of refraction mismatch with the application of special chemicals, and at the same time reduce the scattering coefficient [78]. This method, also known as optical projection tomography, shows very promising results [79]. However, as an in vivo imaging method, major obstacles need to be overcome before this imaging modality reaches practicability.

7.3 Advanced Image Processing

At the conclusion of this chapter, it is also necessary to mention trends and progress in computerized image processing. In the previous chapters, we have seen to what extent image formation depends on the ability of computers to collect and transform data. New image processing methods and higher computing power both work together to provide improved image quality. Most of the backend image processing takes place after the image has been generated: image enhancement, detection (segmentation) of objects of interest, or the measurement of, for example, density or size of a feature of interest. However, as part of the image formation process, improved algorithms play an important role more "behind the curtains". One example was introduced in Chap. 3, where the deconvolution kernel in the filtered backprojection determines the balance between detail and artifactual texture.

Several more examples are introduced below to serve as illustration how progress in computerized image processing influences the entire image formation chain.

After its discovery, the wavelet transform has rapidly found its place in image processing [30]. Donoho and Johnstone proposed powerful wavelet-based noise-reduction methods [80], and many attempts have been made to use wavelet-based denoising most notably in MRI [81], multimodality PET images [82], and ultrasound [83], although ultimately other filter approaches may prove superior [84]. The phenomenon of noise (especially the multiplicative noise in PET and ultrasound) is still subject of active research. Noise reduction is a key element for further progress, because it would allow reconstruction of a tomographic image with less data, which in turn translates into reduced image acquisition times and—in the case of X-ray- or radionuclide-based imaging—reduced radiation exposure.

With a similar goal, that is, image formation from fewer measurements, *compressed sensing* may become the next frontier in tomographic imaging. Compressed sensing was introduced by Candès et al. [85] and is based on the observation that sparse signals can be reconstructed far above the Shannon sampling limit. The principle is taken to the extreme in the model of a single-pixel camera, which takes a limited number of exposures (samples) to form an image of the photographed object. New reconstruction algorithms for CT [86] and MRI [87] have already been introduced.

Another recent development is the more and more prominent role of computer graphics hardware in image processing. The popularity of 3D gaming has made available massively parallel processing engines (known as GPU or *graphics processing units*) [88]. In light of their raw processing power, GPUs are extraordinarily cheap. With GPU support, many vector operations can be accelerated by several orders of magnitude [89]. A typical example is arithmetic reconstruction (ART, cf. Sect. 4.4), where GPU acceleration makes a 1000- to 2000-fold reduction of reconstruction time feasible. Hardware-accelerated versions of the time-consuming cone-beam

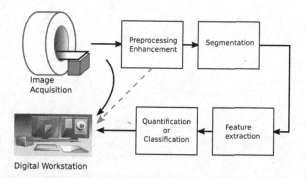

Fig. 7.2 Processing steps in modern computer-aided radiology. Conventionally, the image from the scanner is printed on film or displayed on a digital workstation (*curved arrow*). In computer-aided radiology, the computer performs several steps, starting with image preprocessing and enhancement. The result of this step is valuable in itself and can be provided to the radiologist (*dashed arrow*). For the computer to aid in decision-making, the object of interest needs to be isolated (segmentation), and one or more descriptive metrics extracted (feature extraction). These metrics, or a decision based on them (i.e, classification), are then provided to the radiologist

reconstruction process in CT have also become available [90]. Similarly, hardware-accelerated rigid- or deformable-body registration for multimodality imaging has become available. Clearly, GPU acceleration enables the use of more sophisticated algorithms that would be prohibitively time-consuming on conventional CPUs.

The philosopher's stone of automated image processing is the automated, computerized diagnosis of a disease or anomaly from a given image. This goal is out of reach for any foreseeable future. However, the computer is still capable of aiding the radiologist in various ways, which are often summarized under the term *computer-aided diagnosis* (see e.g., [91, 92] for reviews). The image processing steps in computer-aided diagnosis are highlighted in Fig. 7.2. In conventional radiology, the image is directly displayed or printed on film. Computerized image enhancement, such as noise reduction or adaptive contrast enhancement, is a step that is often included in the digital workstation and is in itself a help for the radiologist. The computer can further support the decision process by (a) extracting the object of interest (segmentation), (b) deriving quantitative descriptors (feature extraction), and (c) proposing a classification (e.g., healthy versus diseased). The final decision lies with the examining radiologist. However, the computer is more and more able to provide objective data and objective measurements to support such a decision. With the development of new methods and the increasing computing power at our disposal, this trend can be expected to continue. Specifically, we can expect ongoing integration of the steps highlighted in Fig. 7.2 into the scanner's image formation software. As such, the radiologist will be more and more able to rely on specific exam modules, for example, *bone density* or *lung emphysema*, that complement the pure image acquisition. As a consequence, the radiologist's decision becomes more objective, can be reached in shorter time, and high-level diagnostic services therefore become more broadly available.

References

1. Bushberg JT, Seibert JA, Leidholdt EM, Boone JM. The essential physics of medical imaging. Philadelphia: Lippincott Williams & Wilkins; 2002.
2. Kalender WA. Computed tomography: fundamentals, system technology, image quality, applications. Erlangen: Publicis; 2011.
3. Kak AC, Slaney M. Principles of computerized tomographic imaging. New York: IEEE Press; 1998. http://www.slaney.org/pct/pct-toc.html. Accessed Aug 2012.
4. Herman GT. Fundamentals of computerized tomography: image reconstruction from projections. Berlin: Springer; 2009.
5. Haacke EM, Brown RW, Thompson MR, Venkatesan R. Magnetic resonance—physical principles and sequence design. New York: Wiley; 1999.
6. Hedrick WR, Hykes DL, Starchman DE. Ultrasound physics and instrumentation. St. Louis: Elsevier Mosby; 2005.
7. Hounsfield GN. Computerized transverse axial scanning (tomography): Part 1. Description of system. Br J Radiol. 1973;46(552):1016–22.
8. Chantler CT, Olsen K, Dragoset RA, Chang J, Kishore AR, Kotochigova SA, et al. Detailed tabulation of atomic form factors, photoelectric absorption and scattering cross section, and mass attenuation coefficients for $Z = 1$–92 from $E = 1$–10 eV to $E = 0.4$–1.0 MeV; 2005. NIST Standard Reference Database 66. http://www.nist.gov/pml/data/ffast/index.cfm. Accessed July 2012.
9. Ho CP, Kim RW, Schaffler MB, Sartoris DJ. Accuracy of dual-energy radiographic absorptiometry of the lumbar spine: cadaver study. Radiology. 1990;176(1):171.
10. Engstrom RW. Photomultiplier handbook. Lancaster: RCA Corp.; 1980. Available on-line at http://psec.uchicago.edu/links/Photomultiplier_Handbook.pdf. Accessed April 2013.
11. Radon J. Über die Bestimmung von Funktionen durch ihre Integralwerte längs gewisser Mannigfaltigkeiten. Ber Sächs Akad Wiss. 1917;69:262–77.
12. Radon J. On the determination of functions from integral values along certain manifolds (translated by P.C. Parks). IEEE Trans Med Imaging. 1986;5(4):170–6.
13. Cormack AM. Representation of a function by its line integrals, with some radiological applications. J Appl Phys. 1963;34(9):2722–7.
14. Cormack AM. Representation of a function by its line integrals, with some radiological applications. II. J Appl Phys. 1964;35(10):2908–13.
15. Ramachandran GN, Lakshminarayanan AV. Three-dimensional reconstruction from radiographs and electron micrographs: application of convolutions instead of Fourier transforms. Proc Natl Acad Sci U S A. 1971;68(9):2236–40.
16. Shepp LA, Logan BF. The Fourier reconstruction of a head section. IEEE Trans Nucl Sci. 1974;21(3):21–43.
17. Rieder A. Principles of reconstruction filter design in 2D-computerized tomography. Contempo Math. 2001;278:207–26.

M. A. Haidekker, *Medical Imaging Technology*, SpringerBriefs in Physics,
DOI: 10.1007/978-1-4614-7073-1, © The Author(s) 2013

18. Feldkamp LA, Davis LC, Kress JW. Practical cone-beam algorithm. J Opt Soc Am A. 1984;1(6):612–9.

19. Grangeat P. Mathematical framework of cone beam 3D reconstruction via the first derivative of the Radon transform. In: Herman G, Luis AK, Natterer F, editors. Mathematical methods in tomography. Berlin: Springer; 1991. p. 66–97.

20. Kyriakou Y, Meyer E, Prell D, Kachelriess M. Empirical beam hardening correction (EBHC) for CT. Med Phys. 2010;37:5179–87.

21. Van Gompel G, Van Slambrouck K, Defrise M, Batenburg KJ, de Mey J, Sijbers J, et al. Iterative correction of beam hardening artifacts in CT. Med Phys. 2011;38 (Suppl.1):S36.

22. Anger HO. Scintillation camera with multichannel collimators. J Nucl Med. 1964;5(7):515–31.

23. Anger HO. Scintillation camera. Rev Sci Instrum. 1958;29(1):27–33.

24. Milster TD, Aarsvold JN, Barrett HH, Landesman AL, Mar LS, Patton DD, et al. A full-field modular gamma camera. J Nucl Med. 1990;31(5):632.

25. Hunter WCJ. Modeling stochastic processes in gamma-ray imaging detectors and evaluation of a multi-anode PMT scintillation camera for use with maximum-likelihood estimation methods; 2007.

26. Kaczmarz S. Angenäherte Auflösung von Systemen linearer Gleichungen [Approximate Solution of Linear Equation Systems]. Bull Int Acad Polon Sci Lett. 1937;A35:355–7.

27. Lange K, Carson R. EM reconstruction algorithms for emission and transmission tomography. J Comput Assist Tomogr. 1984;8(2):306.

28. Green PJ. Bayesian reconstructions from emission tomography data using a modified EM algorithm. IEEE Trans Med Imaging. 1990;9(1):84–93.

29. Judenhofer MS, Wehrl HF, Newport DF, Catana C, Siegel SB, Becker M, et al. Simultaneous PET-MRI: a new approach for functional and morphological imaging. Nat Med. 2008;14(4):459–65.

30. Haidekker MA. Image registration. In: Haidekker MA, Advanced biomedical image analysis. New York: Wiley; 2011.

31. Dussik KT. Über die Möglichkeit, hochfrequente mechanische Schwingungen als diagnostisches Hilfsmittel zu verwerten [On the possibility of using high-frequency mechanical waves as a diagnostic aid]. Z Neurol Psychiat. 1942;174(1):153–68.

32. Donald I. Sonar-the story of an experiment. Ultrasound Med Biol. 1974;1(2):109–17.

33. Donald I. Apologia: how and why medical sonar developed. Ann R Coll Surg Engl. 1974;54(3):132–40.

34. Ikeda TO. Fundamentals of piezoelectricity. Oxford: Oxford University Press; 1990.

35. Hunt JW, Arditi M, Foster FS. Ultrasound transducers for pulse-echo medical imaging. IEEE Trans Biomed Eng. 1983;8:453–81.

36. Dückelmann AM, Kalache KD. Three-dimensional ultrasound in evaluating the fetus. Prenat Diagn. 2010;30(7):631–8.

37. Rizzo G, Pietrolucci M, Aiello E, Mammarella S, Bosi C, Arduini D. The role of three-dimensional ultrasound in the diagnosis of fetal congenital anomalies: a review. Minerva Ginecol. 2011;63(5):401.

38. Li G, Citrin D, Camphausen K, Mueller B, Burman C, Mychalczak B, et al. Advances in 4D medical imaging and 4D radiation therapy. Technol Cancer Res Treat. 2008;7(1):67.

39. Sheikh KH, Smith SW, Ramm Ov, Kisslo J. Real-time, three-dimensional echocardiography: feasibility and initial use. Echocardiography. 1991;8(1):119–25.

40. Hung J, Lang R, Flachskampf F, Shernan SK, McCulloch ML, Adams DB, et al. 3D echocardiography: a review of the current status and future directions. J Am Soc Echocardiogr. 2007;20(3):213–33.

41. Szabo TL. Diagnostic ultrasound imaging: inside out. Burlington: Academic Press; 2004.

42. Kalender WA, Vock P, Polacin A, Soucek M. [Spiral-CT: a new technique for volumetric scans. I. Basic principles and methodology]. Röntgenpraxis; Zeitschrift für Radiologische Technik. 1990;43(9):323.

43. Laskaris ET, Ackermann R, Dorri B, Gross D, Herd K, Minas C. A cryogen-free open superconducting magnet for interventional MRI applications. IEEE Trans Appl Supercond. 1995;5(2):163–8.

44. Ron E. Cancer risks from medical radiation. Health Phys. 2003;85(1):47.

45. Kalender WA, Quick HH. Recent advances in medical physics. Eur Radiol. 2011;21:501–4.

46. Davis TJ, Gao D, Gureyev TE, Stevenson AW, Wilkins SW. Phase-contrast imaging of weakly absorbing materials using hard X-rays. Nature. 1995;373(6515):595–8.

47. Pfeiffer F, Weitkamp T, Bunk O, David C. Phase retrieval and differential phase-contrast imaging with low-brilliance X-ray sources. Nat Phys. 2006;2(4):258–61.

48. Pfeiffer F, Bech M, Bunk O, Kraft P, Eikenberry EF, Brönnimann C, et al. Hard-X-ray dark-field imaging using a grating interferometer. Nat. Mater. 2008;7(2):134–7.

49. Donath T, Pfeiffer F, Bunk O, Grünzweig C, Hempel E, Popescu S, et al. Toward clinical X-ray phase-contrast CT: demonstration of enhanced soft-tissue contrast in human specimen. Invest Radiol. 2010;45(7):445.

50. Bech M, Jensen TH, Bunk O, Donath T, David C, Weitkamp T, et al. Advanced contrast modalities for X-ray radiology: phase-contrast and dark-field imaging using a grating interferometer. Zeitschrift für Medizinische Physik. 2010;20(1):7–16.

51. Webb A. Increasing the sensitivity of magnetic resonance spectroscopy and imaging. Anal Chem. 2012;84(1):9-16.

52. Ogawa S, Lee TM, Kay AR, Tank DW. Brain magnetic resonance imaging with contrast dependent on blood oxygenation. Proc Natl Acad Sci U S A. 1990;87(24):9868.

53. Cercignani M, Horsfield MA. The physical basis of diffusion-weighted MRI. J Neurol Sci. 2001;186:S11–4.

54. Le Bihan D, Mangin JF, Poupon C, Clark CA, Pappata S, Molko N, et al. Diffusion tensor imaging: concepts and applications. J Magn Reson Imaging. 2001;13(4):534–46.

55. Wehrli FW, Saha PK, Gomberg BR, Song HK, Snyder PJ, Benito M, et al. Role of magnetic resonance for assessing structure and function of trabecular bone. Top Magn Reson Imaging. 2002;13(5):335.

56. Magland JF, Wald MJ, Wehrli FW. Spin-echo micro-MRI of trabecular bone using improved 3D fast large-angle spin-echo (FLASE). Magn Reson Med. 2009;61(5):1114–21.

57. Haidekker MA, Dougherty G. In: Dougherty G, editor. Medical imaging in the diagnosis of osteoporosis and estimation of the individual bone fracture risk. Berlin: Springer; 2011. p. 193–225.

58. Prinz C, Voigt JU. Diagnostic accuracy of a hand-held ultrasound scanner in routine patients referred for echocardiography. J Am Soc Echocardiogr. 2011;24(2):111–6.

59. Kiessling F, Fokong S, Koczera P, Lederle W, Lammers T. Ultrasound microbubbles for molecular diagnosis, therapy, and theranostics. J Nucl Med. 2012;53(3):345–8.

60. Ferrara KW, Borden MA, Zhang H. Lipid-shelled vehicles: engineering for ultrasound molecular imaging and drug delivery. Acc Chem Res. 2009;42(7):881–92.

61. Pichler BJ, Wehrl HF, Judenhofer MS. Latest advances in molecular imaging instrumentation. J Nucl Med. 2008;49(Suppl 2):5S–23S.

62. Beyer T, Townsend DW, Brun T, Kinahan PE, Charron M, Roddy R, et al. A combined PET/CT scanner for clinical oncology. J Nucl Med. 2000;41(8):1369–79.

63. Sauter AW, Wehrl HF, Kolb A, Judenhofer MS, Pichler BJ. Combined PET/MRI: one step further in multimodality imaging. Trends Mol Med. 2010;16(11):508–15.

64. Majmudar MD, Nahrendorf M. Cardiovascular molecular imaging: the road ahead. J Nucl Med. 2012;53(5):673–6.

65. Cromer Berman SM, Walczak P, Bulte JWM. Tracking stem cells using magnetic nanoparticles. Wiley Interdiscip Rev: Nanomed Nanobiotechnol. 2011;3:343–55.

66. Wang LV, Hu S. Photoacoustic tomography: in vivo imaging from organelles to organs. Science. 2012;335(6075):1458–62.

67. Geitzenauer W, Hitzenberger CK, Schmidt-Erfurth UM. Retinal optical coherence tomography: past, present and future perspectives. Br J Ophthalmol. 2011;95(2):171–7.

68. Maeda N. Optical coherence tomography for corneal diseases. Eye Contact Lens. 2010;36(5):254.
69. Mogensen M, Thrane L, Jørgensen TM, Andersen PE, Jemec GBE. OCT imaging of skin cancer and other dermatological diseases. J Biophoton. 2009;2:442–51.
70. Osiac E, Säftoiu A, Gheonea DI, Mandrila I, Angelescu R. Optical coherence tomography and Doppler optical coherence tomography in the gastrointestinal tract. World J Gastroenterol. 2011;17(1):15–20.
71. Ripoll J, Nieto-Vesperinas M, Weissleder R, Ntziachristos V. Fast analytical approximation for arbitrary geometries in diffuse optical tomography. Opt Lett. 2002;27(7):527–9.
72. Ntziachristos V, Bremer C, Graves EE, Ripoll J, Weissleder R. In vivo tomographic imaging of near-infrared fluorescent probes. Mol Imaging. 2002;1(2):82–8.
73. Van De Ven S, Elias S, Wiethoff A, Van Der Voort M, Leproux A, Nielsen T, et al. Diffuse optical tomography of the breast: initial validation in benign cysts. Mol Imaging Biol. 2009;11(2):64–70.
74. Choe R, Konecky SD, Corlu A, Lee K, Durduran T, Busch DR, et al. Differentiation of benign and malignant breast tumors by in-vivo three-dimensional parallel-plate diffuse optical tomography. J Biomed Opt. 2009;14:024020.
75. Gladish JC, Yao G, Heureux NL, Haidekker MA. Optical transillumination tomography for imaging of tissue-engineered blood vessels. Ann Biomed Eng. 2005;33(3):323–7.
76. Haidekker MA. Optical transillumination tomography with tolerance against refraction mismatch. Comput Methods Programs Biomed. 2005;80(3):225–35.
77. Huang HM, Xia J, Haidekker MA. Fast optical transillumination tomography with large-size projection acquisition. Ann Biomed Eng. 2008;36(10):1699–707.
78. Sharpe J, Ahlgren U, Perry P, Hill B, Ross A, Hecksher-Sørensen J, et al. Optical projection tomography as a tool for 3D microscopy and gene expression studies. Science. 2002;296(5567):541–5.
79. Swoger J, Sharpe J, Haidekker MA. Optical projection and transillumination tomography for multidimensional mesoscopic imaging. In: Morgan S, Rose FR, Matcher S, editors. Optical techniques in regenerative medicine. New York: Taylor & Francis; 2013.
80. Donoho DL, Johnstone JM. Ideal spatial adaptation by wavelet shrinkage. Biometrika. 1994;81(3):425–55.
81. Zaroubi S, Goelman G. Complex denoising of MR data via wavelet analysis: application for functional MRI. Magn Reson Imaging. 2000;18(1):59–68.
82. Turkheimer FE, Boussion N, Anderson AN, Pavese N, Piccini P, Visvikis D. PET image denoising using a synergistic multiresolution analysis of structural (MRI/CT) and functional datasets. J Nucl Med. 2008;49(4):657–66.
83. Bhuiyan MIH, Omair Ahmad M, Swamy MNS, New spatially adaptive wavelet-based method for the despeckling of medical ultrasound images. In: Circuits and Systems, 2007. ISCAS 2007. IEEE international symposium on. IEEE; 2007. p. 2347–50.
84. Finn S, Glavin M, Jones E. Echocardiographic speckle reduction comparison. IEEE Trans Ultrason Ferroelectr Freq Control. 2011;58(1):82–101.
85. Candès EJ, Romberg J, Tao T. Robust uncertainty principles: exact signal reconstruction from highly incomplete frequency information. IEEE Trans Inf Theory. 2006;52(2):489–509.
86. Li X, Luo S. A compressed sensing-based iterative algorithm for CT reconstruction and its possible application to phase contrast imaging. Biomed Eng Online. 2011;10(1):1–14.
87. Mistretta CA. Undersampled radial MR acquisition and highly constrained back projection (HYPR) reconstruction: potential medical imaging applications in the post-Nyquist era. J Magn Reson Imaging. 2009;29(3):501–16.
88. Etter DM, Hermann RJ. The rise of games and high performance computing for modeling and simulation. Washington, DC: The National Academies Press; 2010.
89. Payne JL, Sinnott-Armstrong NA, Moore JH. Exploiting graphics processing units for computational biology and bioinformatics. Interdiscip Sci: Comput Life Sci. 2010;2(3): 213–20.

90. Zhao X, Hu J, Zhang P. GPU-based 3D cone-beam CT image reconstruction for large data volume. J Biomed Imaging. 2009;2009:8.
91. Erickson BJ, Bartholmai B. Computer-aided detection and diagnosis at the start of the third millennium. J Dig Imaging. 2002;15(2):59–68.
92. Kim TY, Son J, Kim KG. The recent progress in quantitative medical image analysis for computer aided diagnosis systems. Healthcare Inf Research. 2011;17(3):143.

Index

SpringerBriefs in Computer Science

More information about this series at http://www.springer.com/series/10028

Filippo Maria Bianchi · Enrico Maiorino
Michael C. Kampffmeyer
Antonello Rizzi · Robert Jenssen

Recurrent Neural Networks for Short-Term Load Forecasting

An Overview and Comparative Analysis

Springer

Filippo Maria Bianchi
UiT The Arctic University of Norway
Tromsø
Norway

Enrico Maiorino
Harvard Medical School
Boston, MA
USA

Michael C. Kampffmeyer
UiT The Arctic University of Norway
Tromsø
Norway

Antonello Rizzi
Sapienza University of Rome
Rome
Italy

Robert Jenssen
UiT The Arctic University of·Norway
Tromsø
Norway

ISSN 2191-5768   ISSN 2191-5776   (electronic)
SpringerBriefs in Computer Science
ISBN 978-3-319-70337-4   ISBN 978-3-319-70338-1   (eBook)
https://doi.org/10.1007/978-3-319-70338-1

Library of Congress Control Number: 2017957698

Printed on acid-free paper

This Springer imprint is published by Springer Nature
The registered company is Springer International Publishing AG
The registered company address is: Gewerbestrasse 11, 6330 Cham, Switzerland

Preface

The key component in forecasting demand and consumption of resources in a supply network is an accurate prediction of real-valued time series. Indeed, both service interruptions and resource waste can be reduced with the implementation of an effective forecasting system. Significant research has thus been devoted to the design and development of methodologies for short-term load forecasting over the past decades. A class of mathematical models, called recurrent neural networks, are nowadays gaining renewed interest among researchers and they are replacing many practical implementations of the forecasting systems, previously based mostly on statistical methods. Despite the undeniable expressive power of these architectures, their recurrent nature complicates their understanding and poses challenges in the training procedures. Although recently different kinds of recurrent neural networks have been successfully applied in fields like natural language processing or text translation, a systematic evaluation of their performance in the context of load forecasting is still lacking. In this work, we perform a comparative study on the problem of short-term load forecast, by using different classes of state-of-the-art recurrent neural networks. We provide a general overview of the most important architectures and we define guidelines for configuring the recurrent networks to predict real-valued time series. We test the reviewed models on controlled synthetic tasks and on real-world datasets, covering important practical case studies. It is our hope that this essay can become a useful resource for data scientists in academia and industry to keep up-to-date with the latest developments in the field of deep learning and time series prediction.

Tromsø, Norway
September 2017

Filippo Maria Bianchi

Contents

Acronyms

ACEA	Azienda Comunale Energia e Ambiente
ANFIS	Adaptive-Network-Based Fuzzy Inference System
ANN	Artificial Neural Network
ARIMA	Auto-Regressive Integrated Moving Average
BPTT	BackPropagation Through Time
D4D	Data for Development
ERNN	Elman Recurrent Neural Network
ESN	Echo State Network
FFNN	FeedForward Artificial Neural Networks
GEFCom	Global Energy Forecasting Competition
GRU	Gated Recurrent Unit
HF	Hessian Free
LSTM	Long Short-Term Memory
MG	Mackey-Glass
MSE	Mean Square Error
MSO	Multiple Superimposed Oscillators
NARMA	NonLinear Auto-Regressive Moving Average
NARX	Nonlinear Auto-Regressive with eXogenous inputs
NRMSE	Normalized Root Mean Square Error
RNN	Recurrent Neural Network
SGD	Stochastic Gradient Descent
STLF	Short-Term Load Forecast
SVM	Support Vector Machine
TDL	Tapped Delay Line

Chapter 1
Introduction

Abstract A short-term load forecast is the prediction of the consumption of resources in a distribution network in the near future. The supplied resource can be of any kind, such as electricity in power grids or telephone service in telecommunication networks. An accurate forecast of the demand is of utmost importance for the planning of facilities, optimization of day-to-day operations, and an effective management of the available resources. In the context of energy and telecommunication networks, the load data are usually represented as real-valued time series characterized by strong temporal dependencies and seasonal patterns. We begin by reviewing several methods that have been adopted in the past years for the task of short-term load forecast and we highlight their main advantages and limitations. We then introduce the framework of recurrent neural networks, a particular class of artificial neural networks specialized in the processing of sequential/temporal data. We explain how recurrent neural networks can be an effective tool for prediction, especially in those cases where the extent of the time dependencies is unknown a-priori.

Keywords Short-term load forecast · Resource management · Time series prediction · Nonlinear time series analysis · Recurrent neural networks · Predictive models

Forecasting the demand of resources within a distribution network of energy, telecommunication or transportation is of fundamental importance for managing the limited availability of the assets. An accurate Short-Term Load Forecast (STLF) system (Gooijer and Hyndman 2006) can reduce high cost of over- and under-contracts on balancing markets due to load prediction errors. Moreover, it keeps power markets efficient and provides a better understanding of the dynamics of the monitored system (Simchi-Levi et al. 1999). On the other hand, a wrong prediction could cause either a load overestimation, which leads to the excess of supply and consequently more costs and contract curtailments for market participants, or a load underestimation, resulting in failures in gathering enough provisions, thereby more costly supplementary services (Bunn 2000; Ruiz and Gross 2008). These reasons motivated the research of forecasting models capable of reducing this financial distress, by increasing the load

© The Author(s) 2017   1
F. M. Bianchi et al., *Recurrent Neural Networks for Short-Term Load Forecasting*,
SpringerBriefs in Computer Science, https://doi.org/10.1007/978-3-319-70338-1_1

forecasting accuracy even by a small percent (Deihimi and Showkati 2012; Peng et al. 2014; Shen and Huang 2008; Bianchi et al. 2015a, b).

The load profile generally follows cyclic and seasonal patterns related to human activities and can be represented by a real-valued time series. The dynamics of the system generating the load time series can vary significantly during the observation period, depending on the nature of the system and on latent, external influences. For this reason, the forecasting accuracy can change considerably among different samples even when using the same prediction model (Deihimi et al. 2013). Over the past years, the STLF problem has been tackled in several research areas (Jan van Oldenborgh et al. 2005) by means of many different model-based approaches, each one characterized by different advantages and drawbacks in terms of prediction accuracy, complexity in training, sensitivity to the parameters, and limitations in the tractable forecasting horizon (Dang-Ha et al. 2017).

Auto-regressive and exponential smoothing models represented for many years the baseline among systems for time series prediction (Hyndman et al. 2008). Such models require to properly select the lagged inputs to identify the correct model orders, a procedure which demands a certain amount of skill and expertise (Box et al. 2011). Moreover, auto-regressive models make explicit assumptions about the nature of system under analysis. Therefore, their use is limited to those settings in which such assumptions hold and where a-priori knowledge on the system is available (Box and Cox 1964). Taylor (2008) showed that for long forecasting horizons a very basic averaging model, like auto-regressive integrated moving average or triple exponential smoothing, can outperform more sophisticated alternatives. However, in many complicated systems, the properties of linearity and even stationarity of the analyzed time series are not guaranteed. Nonetheless, given their simplicity, auto-regressive models have been largely employed as practical implementations of forecast systems.

The problem of time series prediction has been approached within a function approximation framework, by relying on the embedding procedure proposed by Takens (1981). Takens' theorem transforms the prediction problem from time extrapolation to phase space interpolation. In particular, by properly sampling a time-dependent quantity $s[t]$, it is possible to predict the value of the kth sample from the previous ones, given an appropriate choice of the sampling frequency τ and the number of samples m: $s[k] = f(s[k - \tau], \ldots, s[k - m \cdot \tau])$. Through the application of phase space embedding, regression methods, such as support vector regression (an extension of support vector machines in the continuum), have been applied in time series prediction (Sapankevych and Sankar 2009), either by representing the sequential input as a static domain, described by frequency and phase, or by embedding sequential input values in time windows of fixed length. The approach can only succeed if there are no critical temporal dependencies exceeding the windows length, making the SVM unable to learn an internal state representation for sequence learning tasks involving time lags of arbitrary length. Other universal function approximators such as feedforward artificial neural networks (Hornik et al. 1989) and Adaptive-Network-Based Fuzzy Inference System (ANFIS) (Jang 1993) have been employed in time series prediction tasks by selecting a suitable interval of past values from the time series as the inputs and by training the network to forecast one or a fixed number of future values (Zhang et al. 1998; Hippert et al. 2001; Law 2000; Tsaur et al. 2002; Kon and Turner 2005; Palmer et al. 2006; Claveria and Torra 2014). The operation is repeated to forecast next values by translating the time window of the considered inputs (Kourentzes 2013). While this approach proved to be effective in many circumstances (Daz-Robles et al. 2008; Plummer 2000; Teixeira and Fernandes 2012; Claveria et al. 2015), it does not treat temporal ordering as an explicit feature of the time series and, in general, is not suitable in cases where the time series have significantly different lengths. On this account, a Recurrent Neural Network (RNN) is a more flexible model, since it encodes the temporal context in its feedback connections, which are capable of capturing the time-varying dynamics of the underlying system (Schäfer and Zimmermann 2007; Bianchi et al. 2017).

RNNs are a special class of neural networks characterized by internal self-connections, which can be, in principle, any nonlinear dynamical system, up to a given degree of accuracy (Schäfer and Zimmermann 2007). RNNs and their variants have been used in many contexts where the temporal dependency in the data is an important implicit feature in the model design. Noteworthy applications of RNNs include sequence transduction (Graves 2012), language modeling (Graves 2013; Pascanu et al. 2013b; Mikolov 2012; Sutskever et al. 2011), speech recognition (Graves 2011), learning word embeddings (Mikolov et al. 2013), audio modeling (Oord et al. 2016), handwriting recognition (Graves and Schmidhuber 2009; Graves et al. 2008), and image generation (Gregor et al. 2015). In many of these works, a popular variant of RNN was used, called long short-term memory (Hochreiter and Schmidhuber 1997). This latter has recently earned significant attention due to its capability of storing information for very long periods of time.

As an RNN processes sequential information, it performs the same operations on every element of the input sequence. Its output, at each time step, depends on previous inputs and past computations. This allows the network to develop a memory of previous events, which is implicitly encoded in its hidden state variables. This is certainly different from traditional feedforward neural networks, where it is assumed that all inputs (and outputs) are independent of each other. Theoretically, RNNs can remember arbitrarily long sequences. However, their memory is in practice limited by their finite size and, more critically, by the suboptimal training of their parameters. To overcome memory limitations, recent research efforts have led to the design of novel RNN architectures, which are equipped with an external, permanent memory capable of storing information for indefinitely long amount of time (Weston et al. 2014; Graves et al. 2014).

Contrarily to other linear models adopted for prediction, RNNs can learn functions of arbitrary complexity and they can deal with time series data possessing properties such as saturation or exponential effects and nonlinear interactions between latent variables. However, if the temporal dependencies of data are prevalently contained in a finite and small time interval, the use of RNNs can be unnecessary. In these cases, performances, both in terms of computational resources required and accuracy, are generally lower than the ones of time-window approaches, like ARIMA, SVM, multilayer perceptron, and ANFIS. On the other hand, in many load forecasting problems, the time series to be predicted are characterized by long temporal dependencies, whose extent may vary in time or be unknown in advance. In all these situations, the use of RNNs may turn out to be the best solution.

Despite the STLF problem has been one of the most important applications for both early RNNs models (Vermaak and Botha 1998; Gers et al. 2001) and most recent ones (Flunkert et al. 2017), an up-to-date and comprehensive analysis of the modern RNN architectures applied to the STLF problem is still lacking. In several recent works on STFL, NARX networks (see Sect. 4.1) or Echo State Networks (see Sect. 4.2) are adopted for time series prediction and their performances are usually compared with standard static models, rather than with other RNN architectures. With this overview, we aim to fill these gaps by performing a comparative study on the problem of STLF using different classes of state-of-the-art RNNs. We provide an introduction to the RNN framework, presenting the most important architectures and their properties. We also furnish the guidelines for configuring and training the different RNN models to predict real-valued time series. In practice, we formulate the STLF problem as the prediction of a real-valued univariate time series, given its past values as input. In some cases, beside the time series of past target values, additional "context" time series are fed to the network in order to provide exogenous information related to the environment in which the system to be modeled operates.

This book is structured as follows.

In Chap. 2, we provide a general overview of a standard RNN architecture and we discuss its general properties. We also discuss the main issues encountered in the training phase, the most common methodologies for learning the model parameters and common ways of defining the loss function to be optimized during the training.

In Chap. 3, we present the most basic architecture, called Elman RNN, and then we analyze two important variants, namely the long short-term memory and gated recurrent units networks. Despite the recent popularity of these architectures (Greff et al. 2015), their application to prediction of real-valued time series has been limited so far (Malhotra et al. 2015). For each RNN, we provide a brief review, explaining its main features, the approaches followed in the training stage and a short list of the main works concerning time series prediction in which the specific network has been applied.

Successively, in Chap. 4, we illustrate two particular RNN architectures, which differ from the previous ones, mainly due to their training procedure. In particular, we analyze the Nonlinear Auto-Regressive with eXogenous inputs (NARX) neural network and the Echo State Network (ESN). These architectures have been successfully applied in the literature of time series prediction and they provide important advantages with respect to traditional models, due to their easy applicability and fast training procedures.

In Chap. 5, we describe three synthetic datasets, used to test and compare the computational capabilities of the five RNN architectures in a controlled environment.

In Chap. 6, we present three real-world datasets of time series relative to the load profile in energy distribution and telecommunication networks. For each dataset, we perform a series of analysis with the purpose of choosing a suitable preprocessing for the data.

Chapter 7 is dedicated to the experiments and to the discussion of the performance of the RNN models. The first part of the experimental section focuses on the benchmark tests, while in the second part we employ the RNNs to solve STLF tasks on real-world time series.

Finally, in Chap. 8, we discuss our conclusions.

References

Bianchi FM, De Santis E, Rizzi A, Sadeghian A (2015a) Short-term electric load forecasting using echo state networks and PCA decomposition. IEEE Access 3:1931–1943. doi:https://doi.org/10.1109/ACCESS.2015.2485943

Bianchi FM, Kampffmeyer M, Maiorino E, Jenssen R (2017) Temporal overdrive recurrent neural network. arXiv preprint arXiv:170105159

Bianchi FM, Scardapane S, Uncini A, Rizzi A, Sadeghian A (2015b) Prediction of telephone calls load using echo state network with exogenous variables. Neural Netw 71:204–213. doi:https://doi.org/10.1016/j.neunet.2015.08.010

Box GE, Cox DR (1964) An analysis of transformations. J R Stat Soc Ser B Methodol 211–252

Box GE, Jenkins GM, Reinsel GC (2011) Time series analysis: forecasting and control, vol 734. Wiley

Bunn DW (2000) Forecasting loads and prices in competitive power markets. Proc IEEE 88(2)

Claveria O, Monte E, Torra S (2015) Tourism demand forecasting with neural network models: different ways of treating information. Int J Tour Res 17(5):492–500. doi:https://doi.org/10.1002/jtr.2016,jTR-13-0416.R2

Claveria O, Torra S (2014) Forecasting tourism demand to catalonia: Neural networks vs. time series models. Econo Modell 36:220–228. doi:https://doi.org/10.1016/j.econmod.2013.09.024

Dang-Ha TH, Bianchi FM, Olsson R (2017) Local short term electricity load forecasting: automatic approaches. arXiv:1702:08025

Daz-Robles LA, Ortega JC, Fu JS, Reed GD, Chow JC, Watson JG, Moncada-Herrera JA (2008) A hybrid ARIMA and artificial neural networks model to forecast particulate matter in urban areas: The case of temuco, chile. Atmos Environ 42(35):8331–8340. doi:https://doi.org/10.1016/j.atmosenv.2008.07.020

Deihimi A, Showkati H (2012) Application of echo state networks in short-term electric load forecasting. Energy 39(1):327–340

Deihimi A, Orang O, Showkati H (2013) Short-term electric load and temperature forecasting using wavelet echo state networks with neural reconstruction. Energy 57:382–401

Flunkert V, Salinas D, Gasthaus J (2017) DeepAR: probabilistic forecasting with autoregressive recurrent networks. arXiv:1704:04110

Gers FA, Eck D, Schmidhuber J (2001) Applying lstm to time series predictable through time-window approaches. In: Dorffner G, Bischof H, Hornik K (eds) Artificial Neural Networks—ICANN 2001: 2001 Proceedings International Conference Vienna, Austria, August 21–25. Springer, Berlin, Heidelberg, pp 669–676. doi:https://doi.org/10.1007/3-540-44668-0_93

Gooijer JGD, Hyndman RJ (2006) 25 years of time series forecasting. Int J Forecast 22(3):443–473. doi:https://doi.org/10.1016/j.ijforecast.2006.01.001 (twenty five years of forecasting)

Graves A (2011) Practical variational inference for neural networks. In: Advances in neural information processing systems, pp 2348–2356

Graves A (2012) Sequence transduction with recurrent neural networks. arXiv preprint arXiv:12113711

Graves A (2013) Generating sequences with recurrent neural networks, pp 1–43. arXiv preprint arXiv:13080850, arXiv:1308.0850v5

Graves A, Liwicki M, Bunke H, Schmidhuber J, Fernández S (2008) Unconstrained on-line hand-writing recognition with recurrent neural networks. In: Advances in Neural Information Processing Systems, pp 577–584

Graves A, Schmidhuber J (2009) Offline handwriting recognition with multidimensional recurrent neural networks. In: Advances in neural information processing systems, pp 545–552

Graves A, Wayne G, Danihelka I (2014) Neural turing machines. arXiv:1410.5401

Greff K, Srivastava RK, Koutník J, Steunebrink BR, Schmidhuber J (2015) LSTM: a search space odyssey. arXiv preprint arXiv:150304069

Gregor K, Danihelka I, Graves A, Rezende DJ, Wierstra D (2015) Draw: A recurrent neural network for image generation. arXiv preprint arXiv:150204623

Hippert H, Pedreira C, Souza R (2001) Neural networks for short-term load forecasting: a review and evaluation. IEEE Trans Power Syst 16(1):44–55. doi:https://doi.org/10.1109/59.910780

Hochreiter S, Schmidhuber J (1997) Long short-term memory. Neural Comput 9(8):1735–1780

Hornik K, Stinchcombe M, White H (1989) Multilayer feedforward networks are universal approximators. Neural Netw 2(5):359–366. doi:https://doi.org/10.1016/0893-6080(89)90020-8

Hyndman R, Koehler AB, Ord JK, Snyder RD (2008) Forecasting with exponential smoothing: the state space approach. Springer Science & Business Media

Jan van Oldenborgh G, Balmaseda MA, Ferranti L, Stockdale TN, Anderson DL (2005) Did the ECMWF seasonal forecast model outperform statistical ENSO forecast models over the last 15 years? J Clim 18(16):3240–3249

Jang JSR (1993) Anfis: adaptive-network-based fuzzy inference system. IEEE Trans Syst Man Cybern 23(3):665–685

Kon SC, Turner LW (2005) Neural network forecasting of tourism demand. Tour Econ 11(3):301–328. doi:https://doi.org/10.5367/000000005774353006

Kourentzes N (2013) Intermittent demand forecasts with neural networks. Int J Prod Econ 143(1):198–206. doi:https://doi.org/10.1016/j.ijpe.2013.01.009

Law R (2000) Back-propagation learning in improving the accuracy of neural network-based tourism demand forecasting. Tour Manag 21(4):331–340. doi:https://doi.org/10.1016/S0261-5177(99)00067-9

Malhotra P, Vig L, Shroff G, Agarwal P (2015) Long short term memory networks for anomaly detection in time series. In: Proceedings. Presses universitaires de Louvain, p 89

Mikolov T (2012) Statistical language models based on neural networks. PhD thesis, Brno University of Technology

Mikolov T, Sutskever I, Chen K, Corrado GS, Dean J (2013) Distributed representations of words and phrases and their compositionality. In: Advances in neural information processing systems, pp 3111–3119

Oord Avd, Dieleman S, Zen H, Simonyan K, Vinyals O, Graves A, Kalchbrenner N, Senior A, Kavukcuoglu K (2016) Wavenet: A generative model for raw audio. arXiv preprint arXiv:160903499

Palmer A, Montao JJ, Ses A (2006) Designing an artificial neural network for forecasting tourism time series. Tour Manag 27(5):781–790. doi:https://doi.org/10.1016/j.tourman.2005.05.006

Pascanu R, Mikolov T, Bengio Y (2013b) On the difficulty of training recurrent neural networks. In: Proceedings of the 30th International Conference on International Conference on Machine Learning, JMLR.org, ICML'13, vol 28, pp III–1310–III–1318. http://dl.acm.org/citation.cfm?id=3042817.3043083

Peng Y, Lei M, Li JB, Peng XY (2014) A novel hybridization of echo state networks and multiplicative seasonal ARIMA model for mobile communication traffic series forecasting. Neural Comput Appl 24(3–4):883–890

Plummer E (2000) Time series forecasting with feed-forward neural networks: guidelines and limitations. Neural Netw 1:1

Ruiz PA, Gross G (2008) Short-term resource adequacy in electricity market design. IEEE Trans Power Syst 23(3):916–926

Sapankevych NI, Sankar R (2009) Time series prediction using support vector machines: a survey. IEEE Comput Intell Mag 4(2):24–38

Schäfer AM, Zimmermann HG (2007) Recurrent neural networks are universal approximators. Int J Neural Syst 17(04):253–263. doi:https://doi.org/10.1142/S0129065707001111

Schäfer AM, Zimmermann HG (2007) Recurrent Neural Networks are universal approximators. Int J Neural Syst 17(4):253–263. doi:https://doi.org/10.1142/S0129065707001111

Shen H, Huang JZ (2008) Interday forecasting and intraday updating of call center arrivals. Manuf Serv Oper Manag 10(3):391–410

Simchi-Levi D, Simchi-Levi E, Kaminsky P (1999) Designing and managing the supply chain: concepts, strategies, and cases. McGraw-Hill, New York

Sutskever I, Martens J, Hinton GE (2011) Generating text with recurrent neural networks. In: Proceedings of the 28th International Conference on Machine Learning (ICML-11), pp 1017–1024

Takens F (1981) Detecting strange attractors in turbulence. Springer

Taylor JW (2008) A comparison of univariate time series methods for forecasting intraday arrivals at a call center. Manag Sci 54(2):253–265

Teixeira JP, Fernandes PO (2012) Tourism time series forecast-different ANN architectures with time index input. Procedia Technol 5:445–454. doi:https://doi.org/10.1016/j.protcy.2012.09.049

Tsaur SH, Chiu YC, Huang CH (2002) Determinants of guest loyalty to international tourist hotelsa neural network approach. Tour Manag 23(4):397–405. doi:https://doi.org/10.1016/S0261-5177(01)00097-8

Vermaak J, Botha EC (1998) Recurrent neural networks for short-term load forecasting. IEEE Trans Power Syst 13(1):126–132. https://doi.org/10.1109/59.651623

Weston J, Chopra S, Bordes A (2014) Memory networks. arXiv:1410.3916

Zhang G, Patuwo BE, Hu MY (1998) Forecasting with artificial neural networks:: the state of the art. Int J Forecast 14(1):35–62. doi:https://doi.org/10.1016/S0169-2070(97)00044-7

Chapter 2
Properties and Training in Recurrent Neural Networks

Abstract In this chapter, we describe the basic concepts behind the functioning of recurrent neural networks and explain the general properties that are common to several existing architectures. We introduce the basis of their training procedure, the backpropagation through time, as a general way to propagate and distribute the prediction error to previous states of the network. The learning procedure consists of updating the model parameters by minimizing a suitable loss function, which includes the error achieved on the target task and, usually, also one or more regularization terms. We then discuss several ways of regularizing the system, highlighting their advantages and drawbacks. Beside the standard stochastic gradient descent procedure, we also present several additional optimization strategies proposed in the literature for updating the network weights. Finally, we illustrate the problem of the vanishing gradient effect, an inherent problem of the gradient-based optimization techniques which occur in several situations while training neural networks. We conclude by discussing the most recent and successful approaches proposed in the literature to limit the vanishing of the gradients.

Keywords Learning procedures in neural networks · Parameters training · Gradient descent · Backpropagation through time · Loss function · Regularization techniques · Vanishing gradient

RNNs are learning machines that recursively compute new states by applying transfer functions to previous states and inputs. Typical transfer functions are composed by an affine transformation followed by a nonlinear function, which are chosen depending on the nature of the particular problem at hand. It has been shown by Maass et al. (2007) that RNNs possess the so-called universal approximation property, that is, they are capable of approximating arbitrary nonlinear dynamical systems (under loose regularity conditions) with arbitrary precision, by realizing complex mappings from input sequences to output sequences Siegelmann and Sontag (1991). However, the particular architecture of an RNN determines how information flows between different neurons and its correct design is crucial for the realization of a robust learning system. In the context of prediction, an RNN is trained on input temporal data $\mathbf{x}(\mathbf{t})$ in order to reproduce a desired temporal output $\mathbf{y}(\mathbf{t})$. $\mathbf{y}(t)$ can be any time series related to the input and even a temporal shift of $\mathbf{x}(\mathbf{t})$ itself. The most common training procedures are gradient-based, but other techniques have been proposed, based on

© The Author(s) 2017   9
F. M. Bianchi et al., *Recurrent Neural Networks for Short-Term Load Forecasting*,
SpringerBriefs in Computer Science, https://doi.org/10.1007/978-3-319-70338-1_2

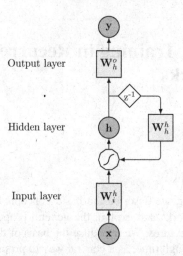

Fig. 2.1 Schematic depiction of a simple RNN architecture. The circles represent input $\mathbf{x}$, hidden, $\mathbf{h}$, and output nodes, $\mathbf{y}$, respectively. The solid squares $\mathbf{W}_i^h$, $\mathbf{W}_h^h$, and $\mathbf{W}_h^o$ are the matrices which represent input, hidden, and output weights respectively. Their values are commonly tuned in the training phase through gradient descent. The polygon represents the nonlinear transformation performed by neurons and z^{-1} is the unit delay operator

derivative-free approaches or convex optimization Schmidhuber et al. (2007), Jaeger (2001). The objective function to be minimized is a loss function, which depends on the error between the estimated output $\hat{\mathbf{y}}(t)$ and the actual output of the network $\mathbf{y}(\mathbf{t})$. An interesting aspect of RNNs is that, upon suitable training, they can also be executed in generative mode, as they are capable of reproducing temporal patterns similar to those they have been trained on Gregor et al. (2015).

The architecture of a simple RNN is depicted in Fig. 2.1. In its most general form, an RNN can be seen as a weighted, directed, and cyclic graph that contains three different kinds of nodes, namely the input, hidden, and output nodes (Zhang et al. 2016). Input nodes do not have incoming connections, output nodes do not have outgoing connections, hidden nodes have both. An edge can connect two different nodes which are at the same or at different time instant. In the following, we adopt the time-shift operator z^n to represent a time delay of n time steps between a source and a destination node. Usually $n = -1$, but also lower values are admitted and they represent the so- called skip connections (Koutník et al. 2014). Self-connecting edges always implement a lag operator with $|n| \geq 1$. In some particular cases, the argument of the time-shift operator is positive and it represents a forward shift in time (Sutskever and Hinton 2010). This means that a node receives as input the content of a source node in a future time interval. Networks with those kind of connections are called bidirectional RNNs and are based on the idea that the output at a given time may not only depend on the previous elements in the sequence, but also on future ones (Schuster and Paliwal 1997). These architectures, however, are not reviewed in this work as we only focus on RNNs with $n = -1$.

While, in theory, an RNN architecture can model any given dynamical system, practical problems arise during the training procedure, when model parameters must be learned from data in order to solve a target task. Part of the difficulty is due to a lack of well-established methodologies for training different types of models. This is also because a general theory that might guide designer decisions has lagged behind the feverish pace of novel architecture designs (Schoenholz et al. 2016; Lipton 2015). A large variety of novel strategies and heuristics have arisen from the literature in the past years (Montavon et al. 2012; Scardapane et al. 2017) and, in many cases, they may require a considerable amount of expertise from the user to be correctly applied. While the standard learning procedure is based on gradient optimization, in some RNN architectures the weights are trained following different approaches (Scardapane and Wang 2017; Jaeger 2002b), such as real-time recurrent learning (Williams and Zipser 1989), extended Kalman filters (Haykin et al. 2001), or evolutionary algorithms (John 1992), and in some cases they are not learned at all (Lukoševičius and Jaeger 2009).

2.1 Backpropagation Through Time

Gradient-based learning requires a closed-form relation between the model parameters and the loss function. This relation allows to propagate the gradient information calculated on the loss function back to the model parameters, in order to modify them accordingly. While this operation is straightforward in models represented by a directed acyclic graph, such as a FeedForward Neural Network (FFNN), some caution must be taken when this reasoning is applied to RNNs, whose corresponding graph is cyclic. Indeed, in order to find a direct relation between the loss function and the network weights, the RNN has to be represented as an equivalent infinite, acyclic, and directed graph. The procedure is called *unfolding* and consists of replicating the network's hidden layer structure for each time interval, obtaining a particular kind of FFNN. The key difference of an unfolded RNN with respect to a standard FFNN is that the weight matrices are constrained to assume the same values in all replicas of the layers, since they represent the recursive application of the same operation.

Figure 2.2 depicts the unfolding of the RNN, previously reported in Fig. 2.1. Through this transformation the network can be trained with standard learning algorithms, originally conceived for feedforward architectures. This learning procedure is called Backpropagation Through Time (BPTT) (Rumelhart et al. 1985) and is one of the most successful techniques adopted for training RNNs. However, while the network structure could in principle be replicated an infinite number of times, in practice the unfolding is always truncated after a finite number of time instants. This maintains the complexity (depth) of the network treatable and limits the issue of the vanishing gradient (as discussed later). In this learning procedure called truncated BPPT (Williams and Peng 1990), the folded architecture is repeated up to a given number of steps τ_b, with τ_b upperbounded by the time series length T. The size of the truncation depends on the available computational resources, as the

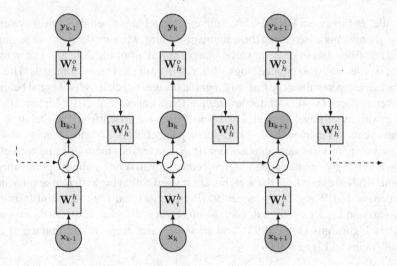

Fig. 2.2 The diagram depicts the RNN from Fig. 2.1, being unfolded (or unrolled) into a FFNN. As we can see from the image, each input $\mathbf{x}_t$ and output $\mathbf{y}_t$ are relative to different time intervals. Unlike a traditional deep FFNN, which uses different parameters in each layer, an unfolded RNN shares the same weights across every time step. In fact, the input weights matrix $\mathbf{W}_i^h$, the hidden weights matrix $\mathbf{W}_h^h$, and the output weights matrix $\mathbf{W}_h^o$ are constrained to keep the same values in each time interval

network grows deeper by repeating the unfolding, and on the expected maximum extent of time dependencies in data. For example, in a periodic time series with period t it may be unnecessary, or even detrimental, to set $\tau_b > t$.

Another variable we consider is the frequency τ_f at which the BPTT calculates the backpropagated gradients. In particular, let us define with BPTT(τ_b, τ_f) the truncated backpropagation that processes the sequence one time step at a time, and every τ_f time steps, it runs BPTT for τ_b time steps (Sutskever 2013). Very often the term τ_f is omitted in the literature, as it is assumed equal to 1, and only the value for τ_b is specified. We refer to the case $\tau_f = 1$ and $\tau_b = n$ as *true* BPTT, or BPTT$(n, 1)$.

In order to improve the computational efficiency of the BPTT, the ratio τ_b/τ_f can be decremented, effectively reducing the frequency of gradients evaluation. An example, is the so-called *epochwise* BPTT or BPTT(n, n), where $\tau_b = \tau_f$ (Williams and Zipser 1995). In this case, the ratio $\tau_b/\tau_f = 1$. However, the learning procedure is in general much less accurate than BPTT$(n, 1)$, since the gradient is truncated too early for many values on the boundary of the backpropagation window.

A better approximation of the true BPTT is reached by taking a large difference $\tau_b - \tau_f$, since no error in the gradient is injected for the earliest $\tau_b - \tau_f$ time steps in the buffer. A good tradeoff between accuracy and performance is BPTT$(2n, n)$, which keeps the ratio $\tau_b/\tau_f = 2$ sufficiently close to 1 and the difference $\tau_b - \tau_f = n$ is large as in the true BPTT (Williams and Peng 1990). Through preliminary experiments, we observed that BPTT$(2n, n)$ achieves comparable performance to BPTT$(n, 1)$, in a significantly reduced training time. Therefore, we followed this procedure in all our experiments.

2.2 Gradient Descent and Loss Function

Training a neural network commonly consists of modifying its parameters through a gradient descent optimization, which minimizes a given loss function that quantifies the accuracy of the network in performing the desired task. The gradient descent procedure consists of repeating two basic steps until convergence is reached. First, the loss function L_k is evaluated on the RNN configured with weights $\mathbf{W}_k$, when a set of input data $\mathcal{X}_k$ are processed (forward pass). Note that with $\mathbf{W}_k$ we refer to *all* network parameters, while the index k identifies their values at epoch k, as they are updated during the optimization procedure. In the second step, the gradient $\partial L_k / \partial \mathbf{W}_k$ is backpropagated through the network in order to update its parameters (backward pass).

In a time series prediction problem, the loss function evaluates the dissimilarity between the predicted values and the actual future values of the time series, which is the ground truth. The loss function can be defined as

$$L_k = E\left(\mathcal{X}_k, \mathcal{Y}_k^*; \mathbf{W}_k\right) + R_\lambda\left(\mathbf{W}_k\right), \tag{2.1}$$

where E is a function that evaluates the prediction error of the network when it is fed with inputs in $\mathcal{X}_k$, in respect to a desired response $\mathcal{Y}_k^*$. R_λ is a regularization function that depends on a hyperparameter λ, which weights the contribution of the regularization in the total loss.

The error function E that we adopt in this work is Mean Square Error (MSE). It is defined as

$$\text{MSE}(\mathcal{Y}_k, \mathcal{Y}_k^*) = \frac{1}{|\mathcal{X}_k|} \sum_{\mathbf{x} \in \mathcal{X}_k} \left(\mathbf{y}_\mathbf{x} - \mathbf{y}_\mathbf{x}^*\right)^2, \tag{2.2}$$

where $\mathbf{y}_\mathbf{x} \in \mathcal{Y}_k$ is the output of the RNN (configured with parameters $\mathbf{W}_k$) when the input $\mathbf{x} \in \mathcal{X}_k$ is processed and $\mathbf{y}_\mathbf{x}^* \in \mathcal{Y}_k^*$ is the ground-truth value that the network must learn to reproduce.

The regularization term R_λ introduces a bias that improves the generalization capabilities of the RNN, by reducing overfitting on the training data. In this work, we consider four types of regularization:

1. L_1: the regularization term in Eq. 2.1 has the form $R_\lambda(\mathbf{W}_k) = \lambda_1 \|\mathbf{W}_k\|_1$. L_1 regularization enforces sparsity in the network parameters, is robust to noisy outliers and it can possibly deliver multiple optimal solutions. However, this regularization can produce unstable results, in the sense that a small variation in the training data can yield very different outcomes.

2. L_2: in this case, $R_\lambda(\mathbf{W}_k) = \lambda_2 \|\mathbf{W_k}\|_2$. This function penalizes large magnitudes in the parameters, favoring dense weight matrices with low values. This procedure is more sensitive to outliers, but is more stable than L_1. Usually, if one is not concerned with explicit features selection, the use of L_2 is preferred.

3. *Elastic net penalty*: combines the two regularizations above, by joining both L_1 and L_2 terms as $R_\lambda(\mathbf{W}_k) = \lambda_1 \|\mathbf{W}_k\|_1 + \lambda_2 \|\mathbf{W}_k\|_2$. This regularization method overcomes the shortcomings of the L_1 regularization, which selects a limited number of variables before it saturates and, in case of highly correlated variables, tends to pick only one and ignore the others. Elastic net penalty generalizes the L_1 and L_2 regularization, which can be obtained by setting $\lambda_2 = 0$ and $\lambda_1 = 0$, respectively.

4. *Dropout*: rather than defining an explicit regularization function $R_\lambda(\cdot)$, dropout is implemented by keeping a neuron active during each forward pass in the training phase with some probability. Specifically, one applies a randomly generated mask to the output of the neurons in the hidden layer. The probability of each mask element to be 0 or 1 is defined by a hyperparameter p_{drop}. Once the training is over, the activations are scaled by p_{drop} in order to maintain the same expected output. Contrary to feedforward architectures, a naive dropout in recurrent layers generally produces bad performance and, therefore, it has usually been applied only to input and output layers of the RNN (Pham et al. 2014a). However, in a recent work Gal and Ghahramani (2015), it is shown that this shortcoming can be circumvented by dropping the same network units in each epoch of the gradient descent. Even if this formulation yields a slightly reduced regularization, nowadays this approach is becoming popular (Zilly et al. 2016; Che et al. 2016) and is the one we followed in our experiments.

Beside the ones discussed above, several other kinds of regularization procedures have been proposed in the literature. Examples are the stochastic noise injection (Neelakantan et al. 2015) and the max-norm constraint (Lee et al. 2010), which, however, are not considered in our experiments.

2.3 Parameters Update Strategies

Rather than evaluating the loss function over the entire training set to perform a single update of the network parameters, a very common approach consists of computing the gradient over mini-batches $\mathscr{X}_k$ of the training data. The size of the batch is usually set by following rules of thumb (Bengio 2012).

This gradient-update method is called Stochastic Gradient Descent (SGD) and, in presence of a non-convex function, its convergence to a local minimum is guaranteed (under some mild assumptions) if the learning rate is sufficiently small (Bottou 2004). The update equation reads

$$\mathbf{W}_{k+1} = \mathbf{W}_k + \eta \nabla L_k(\mathbf{W}_k), \tag{2.3}$$

where η is the *learning rate*, an important hyperparameter that must be carefully tuned to achieve an effective training (Bottou 2012a). In fact, a large learning rate provides a high amount of kinetic energy in the gradient descent, which causes the parameter vector to bounce, preventing the access to narrow area of the search space, where the loss function is lower. On the other hand, a strong decay can excessively slow the training procedure, resulting in a waste of computational time.

Several solutions have been proposed over the years, to improve the convergence to the optimal solution (Bottou 2012b). During the training phase, it is usually helpful to anneal η over time or when the performance stops increasing. A method called *step decay* reduces the learning rate by a factor α, if after a given number of epochs the loss has not decreased. The *exponential decay* and the *fractional decay* instead, have mathematical forms $\eta = \eta_0 e^{-\alpha k}$ and $\eta = \frac{\eta_0}{(1+\alpha k)}$, respectively. Here α and η_0 are hyperparameters, while k is the current optimization epoch. In our experiments, we opted for the step decay annealing, when we train the networks with SGD.

Even if SGD usually represents a safe optimization procedure, its rate of convergence is slow and the gradient descent is likely to get stuck in a saddle point of the loss function landscape (Dauphin et al. 2014). Those issues have been addressed by several alternative strategies proposed in the literature for updating the network parameters. In the following, we describe the most commonly used ones.

Momentum

In this first-order method, the weights $\mathbf{W}_k$ are updated according to a linear combination of the current gradient $\nabla L_k(\mathbf{W}_k)$ and the previous update $\mathbf{V}_{k-1}$, which is scaled by a hyperparameter μ:

$$\begin{aligned}
\mathbf{V}_k &= \mu \mathbf{V}_{k-1} - \eta \nabla L_k(\mathbf{W}_k), \\
\mathbf{W}_{k+1} &= \mathbf{W}_k + \mathbf{V}_k.
\end{aligned} \tag{2.4}$$

With this approach, the updates will build up velocity toward a direction that shows a consistent gradient (Sutskever 2013). A common choice is to set $\mu = 0.9$.

A variant of the original formulation is the *Nesterov momentum*, which often achieves a better convergence rate, especially for smoother loss functions (Nesterov 1983). Contrary to the original momentum, the gradient is evaluated at an approximated future location, rather than at the current position. The update equations are

$$\begin{aligned}
\mathbf{V}_k &= \mu \mathbf{V}_{k-1} - \eta \nabla L_k(\mathbf{W}_k + \mu \mathbf{V}_{k-1}), \\
\mathbf{W}_{k+1} &= \mathbf{W}_k + \mathbf{V}_k.
\end{aligned} \tag{2.5}$$

Adaptive Learning Rate

The first adaptive learning rate method, proposed by Duchi et al. (2011), is Adagrad. Unlike the previously discussed approaches, Adagrad maintains a different learning rate for each parameter. Given the update information from all previous iterations $\nabla L_k(\mathbf{W}_j)$, with $j \in \{0, 1, \cdots, k\}$, a different update is specified for each parameter i of the weight matrix:

$$\mathbf{W}_{k+1}^{(i)} = \mathbf{W}_k^{(i)} - \eta \frac{\nabla L_k \left(\mathbf{W}_k^{(i)}\right)}{\sqrt{\sum_{j=0}^{k} \nabla L_k \left(\mathbf{W}_j^{(i)}\right)^2 + \epsilon}}, \qquad (2.6)$$

where ε is a small term used to avoid division by 0. A major drawback with Adagrad is the unconstrained growth of the accumulated gradients over time. This can cause diminishing learning rates that may stop the gradient descent prematurely.

A procedure called RMSprop (Tieleman and Hinton 2012) attempts to solve this issue by using an exponential decaying average of square gradients, which discourages an excessive shrinkage of the learning rates:

$$v_k^{(i)} = \begin{cases} (1 - \delta) \cdot v_{k-1}^{(i)} + \delta \nabla L_k \left(\mathbf{W}_k^{(i)}\right)^2 & \text{if } \nabla L_k \left(\mathbf{W}_k^{(i)}\right) > 0 \\ (1 - \delta) \cdot v_{k-1}^{(i)} & \text{otherwise} \end{cases} \qquad (2.7)$$

$$\mathbf{W}_{k+1}^{(i)} = \mathbf{W}_k^{(i)} - \eta v_k^{(i)}.$$

According to the update formula, if there are oscillation in gradient updates, the learning rate is reduced by $1 - \delta$, otherwise it is increased by δ. Usually the decay rate is set to $\delta = 0.01$.

Another approach called Adam and proposed by Kingma and Ba (2014), combines the principles of Adagrad and momentum update strategies. Usually, Adam is the adaptive learning method that yields better results and, therefore, it is the gradient descent strategy most used in practice. Like RMSprop, Adam stores an exponentially decaying average of gradients squared, but it also keeps an exponentially decaying average of the moments of the gradients. The update difference equations of Adam are as follows:

$$m_k = \beta_1 m_{k-1} + (1 - \beta_1) \nabla L_k \left(\mathbf{W}_k^{(i)}\right),$$

$$v_k = \beta_2 v_{k-1} + (1 - \beta_2) \nabla L_k \left(\mathbf{W}_k^{(i)}\right)^2,$$

$$\hat{m}_k = \frac{m_k}{1 - \beta_1^k}, \quad \hat{v}_k = \frac{v_k}{1 - \beta_2^k}, \qquad (2.8)$$

$$\mathbf{W}_{k+1} = \mathbf{W}_k + \frac{\eta}{\sqrt{\hat{v}_k} + \epsilon} \hat{m}_k.$$

m corresponds to the first moment and v is the second moment. However, since both m and v are initialized as zero-vectors, they are biased toward 0 during the first epochs. To avoid this effect, the two terms are corrected as $\hat{m}_t$ and $\hat{v}_t$. Default values of the hyperparameters are $\beta_1 = 0.9$, $\beta_2 = 0.999$ and $\varepsilon = 10^{-8}$.

Second-order Methods

The methods discussed so far only consider first-order derivatives of the loss function. Due to this approximation, the landscape of the loss function locally looks and behaves like a plane. Ignoring the curvature of the surface may lead the optimization astray and it could cause the training to progress very slowly. However, second-order methods involve the computation of the Hessian, which is expensive and usually untreatable even in networks of medium size. A Hessian-Free (HF) method that considers derivatives of the second order, without explicitly computing the Hessian, has been proposed by Martens (2010). This latter, unlike other existing HF methods, makes use of the positive semi-definite Gauss–Newton curvature matrix and it introduces a damping factor based on the Levenberg–Marquardt heuristic, which permits to train networks more effectively. However, Sutskever et al. (2013) showed that HF obtains similar performance to SGD with Nesterov momentum. Despite being a first-order approach, Nestrov momentum is capable of accelerating directions of low-curvature just like a HF method and, therefore, is preferred due to its lower computational complexity.

2.4 Vanishing and Exploding Gradient

Increasing the depth in an RNN, in general, improves the memory capacity of the network and its modeling capabilities (Pascanu et al. 2013a). For example, stacked RNNs do outperform shallow ones with the same hidden size on problems where it is necessary to store more information throughout the hidden states between the input and output layer (Sutskever et al. 2014). One of the principal drawback of early RNN architectures was their limited memory capacity, caused by the *vanishing* or *exploding gradient* problem (El Hihi and Bengio 1995), which becomes evident when the information contained in past inputs must be retrieved after a long time interval (Hochreiter et al. 2001). To illustrate the issue of vanishing gradient, one can consider the influence of the loss function L_t (that depends on the network inputs and on its parameters) on the network parameters $\mathbf{W}_t$, when its gradient is backpropagated through the unfolded The network Jacobian reads as

$$\frac{\partial L[t]}{\partial W} = \sum_\tau \frac{\partial L[t]}{\partial h[t]} \frac{\partial h[t]}{\partial h[\tau]} \frac{\partial h[\tau]}{\partial W}. \tag{2.9}$$

In the previous equation, the partial derivatives of the states with respect to their previous values can be factorized as

$$\frac{\partial h[t]}{\partial h[\tau]} = \frac{\partial h[t]}{\partial h[t-1]} \cdots \frac{\partial h[\tau+1]}{\partial h[\tau]} = f'_t \cdots f'_{\tau+1}. \tag{2.10}$$

To ensure local stability, the network must operate in a ordered regime (Bianchi et al. 2016a), a property ensured by the condition $|f'_t| < 1$. However, in this case the product expanded in Eq. 2.10 rapidly (exponentially) converges to 0, when $t - \tau$ increases. Consequently, the sum in Eq. 2.9 becomes dominated by terms corresponding to short-term dependencies and the vanishing gradient effect occurs. As a principal side effect, the weights are less and less updated as the gradient flows backward through the layers of the network. On the other hand, the phenomenon of exploding gradient appears when $|f'_t| > 1$ and the network becomes locally unstable.

Even if global stability can still be obtained under certain conditions, in general the network enters into a chaotic regime, where its computational capability is hindered (Livi et al. 2017).

Models with large recurrent depths exacerbate these gradient-related issues, since they posses more nonlinearities and the gradients are more likely to explode or vanish. A common way to handle the exploding gradient problem, is to clip the norm of the gradient if it grows above a certain threshold. This procedure relies on the assumption that exploding gradients only occur in contained regions of the parameters space. Therefore, clipping avoids extreme parameter changes without overturning the general descent direction (Pascanu et al. 2012).

On the other hand, different solutions have been proposed to tackle the vanishing gradient issue. A simple, yet effective approach consists of initializing the weights to maintain the same variance withing the activations and backpropagated gradients, as one moves along the network depth. This is obtained with a random initialization that guarantees the variance of the components of the weight matrix in layer l to be $\mathrm{Var}(\mathbf{W}_l) = 2/(N_{l-1}+N_{l+1})$, N_{l-1} and N_{l+1} being the number of units in the previous and the next layer respectively (Glorot and Bengio 2010). He et al. (2015) proposed to initialize the network weights by sampling them from an uniform distribution in $[0, 1]$ and then rescaling their values by $1/\sqrt{N_h}$, N_h being the total number of hidden neurons in the network. Another option, popular in deep FFNN, consists of using ReLU (Nair and Hinton 2010) as activation function, whose derivative is 0 or 1, and it does not cause the gradient to vanish or explode. Regularization, besides preventing unwanted overfitting in the training phase, proved to be useful in dealing with exploding gradients. In particular, L_1 and L_2 regularizations constrain the growth of the components of the weight matrices and consequently limit the values assumed by the propagated gradient Pascanu et al. (2013b). Another popular solution is adopting gated architectures, like long short-term memory (LSTM) or Gated Recurrent Unit (GRU), which have been specifically designed to deal with vanishing gradients and allow the network to learn much longer range dependencies. Srivastava et al. (2015) proposed an architecture called *Highway Network*, which allows information to flow across several layers without attenuation. Each layer can smoothly vary its behavior between that of a plain layer, implementing an affine transform followed by a nonlinear activation, and that of a layer which simply passes its input through. Optimization in highway networks is virtually independent of depth, as information can be routed (unchanged) through the layers. The highway architecture, initially applied to deep FFNN (He et al. 2015), has recently been extended to RNN where it dealt with several modeling and optimization issues (Zilly et al. 2016).

Finally, gradient-related problems can be avoided by repeatedly selecting new weight parameters using random guess or evolutionary approaches (John 1992; Gomez and Miikkulainen 2003); in this way the network is less likely to get stuck in local minima. However, convergence time of these procedures is time consuming and can be impractical in many real-world applications. A solution proposed by Schmidhuber et al. (2007), consists of evolving only the weights of nonlinear hidden units, while linear mappings from hidden to output units are tuned using fast algorithms for convex problem optimization.

References

Bengio Y (2012) Practical recommendations for gradient-based training of deep architectures. In: Montavon G, Orr GB, Müller KR (eds) Neural networks: tricks of the trade: second edition. Springer, Berlin, pp 437–478. https://doi.org/10.1007/978-3-642-35289-8_26

Bianchi FM, Livi L, Alippi C (2016a) Investigating echo-state networks dynamics by means of recurrence analysis. IEEE Trans Neural Netw Learn Syst 99:1–13. doi:https://doi.org/10.1109/TNNLS.2016.2630802

Bottou L (2004) Stochastic learning. In: Bousquet O, von Luxburg U (eds) Advanced lectures on machine learning. Lecture Notes in Artificial Intelligence, LNAI, vol 3176. Springer Verlag, Berlin, pp 146–168. http://leon.bottou.org/papers/bottou-mlss-2004

Bottou L (2012a) Stochastic gradient descent tricks. In: Neural networks: tricks of the trade. Springer, pp 421–436

Bottou L (2012b) Stochastic gradient descent tricks. In: Montavon G, Orr GB, Müller KR (eds) neural networks: tricks of the trade: second edition. Springer, Berlin, pp 421–436. https://doi.org/10.1007/978-3-642-35289-8_25

Che Z, Purushotham S, Cho K, Sontag D, Liu Y (2016) Recurrent neural networks for multivariate time series with missing values. arXiv:1606.01865

Dauphin YN, Pascanu R, Gulcehre C, Cho K, Ganguli S, Bengio Y (2014) Identifying and attacking the saddle point problem in high-dimensional non-convex optimization. In: Ghahramani Z, Welling M, Cortes C, Lawrence ND, Weinberger KQ (eds) Advances in neural information processing systems, vol 27. Curran Associates Inc., pp 2933–2941

Duchi J, Hazan E, Singer Y (2011) Adaptive subgradient methods for online learning and stochastic optimization. J Mach Learn Res 12:2121–2159

El Hihi S, Bengio Y (1995) Hierarchical recurrent neural networks for long-term dependencies. In: Proceedings of the 8th International Conference on Neural Information Processing Systems (NIPS'95). MIT Press, Cambridge, MA, USA, pp 493–499. http://dl.acm.org/citation.cfm?id=2998828.2998898

Gal Y, Ghahramani Z (2015) A theoretically grounded application of dropout in recurrent neural networks. arXiv:1512:05287

Glorot X, Bengio Y (2010) Understanding the difficulty of training deep feedforward neural networks. In: International conference on artificial intelligence and statistics, pp 249–256

Gomez FJ, Miikkulainen R (2003) Robust non-linear control through neuroevolution. Computer Science Department, University of Texas at Austin

Gregor K, Danihelka I, Graves A, Rezende DJ, Wierstra D (2015) Draw: a recurrent neural network for image generation. arXiv:150204623

Haykin SS, Haykin SS, Haykin SS (2001) Kalman filtering and neural networks. Wiley Online Library

He K, Zhang X, Ren S, Sun J (2015) Deep residual learning for image recognition. arXiv:1512:03385

He K, Zhang X, Ren S, Sun J (2015) Delving deep into rectifiers: Surpassing human-level performance on imagenet classification. In: Proceedings of the IEEE international conference on computer vision, pp 1026–1034

Hochreiter S, Bengio Y, Frasconi P, Schmidhuber J (2001) Gradient flow in recurrent nets: the difficulty of learning long-term dependencies

Jaeger H (2001) The "echo state" approach to analysing and training recurrent neural networks-with an erratum note. German National Research Center for Information Technology GMD Technical Report 148:34, Bonn, Germany

Jaeger H (2002b) Tutorial on training recurrent neural networks, covering BPTT, RTRL, EKF and the "echo state network" approach, vol 5. GMD-Forschungszentrum Informationstechnik

John H (1992) Holland, adaptation in natural and artificial systems

Kingma D, Ba J (2014) Adam: a method for stochastic optimization. arXiv:14126980

Koutník J, Greff K, Gomez FJ, Schmidhuber J (2014) A clockwork RNN. arXiv:1402.3511

Lee JD, Recht B, Srebro N, Tropp J, Salakhutdinov RR (2010) Practical large-scale optimization for max-norm regularization. In: Advances in neural information processing systems, pp 1297–1305

Lipton ZC (2015) A critical review of recurrent neural networks for sequence learning. http://arxiv.org/abs/1506.00019

Livi L, Bianchi FM, Alippi C (2017) Determination of the edge of criticality in echo state networks through fisher information maximization. IEEE Trans Neural Netw Learn Syst (99):1–12. doi:https://doi.org/10.1109/TNNLS.2016.2644268

Lukoševičius M, Jaeger H (2009) Reservoir computing approaches to recurrent neural network training. Comput Sci Rev 3(3):127–149. https://doi.org/10.1016/j.cosrev.2009.03.005

Maass W, Joshi P, Sontag ED (2007) Computational aspects of feedback in neural circuits. PLoS Comput Biol 3(1):e165. https://doi.org/10.1371/journal.pcbi.0020165.eor

Martens J (2010) Deep learning via hessian-free optimization. In: Proceedings of the 27th International Conference on Machine Learning (ICML-10), pp 735–742

Montavon G, Orr G, Müller KR (2012) Neural networks-tricks of the trade second edition. Springer. https://doi.org/10.1007/978-3-642-35289-8

Nair V, Hinton GE (2010) Rectified linear units improve restricted boltzmann machines. In: Proceedings of the 27th International Conference on Machine Learning (ICML-10), June 21-24, 2010, Haifa, Israel, pp 807–814

Neelakantan A, Vilnis L, Le QV, Sutskever I, Kaiser L, Kurach K, Martens J (2015) Adding gradient noise improves learning for very deep networks. arXiv:151106807

Nesterov Y (1983) A method of solving a convex programming problem with convergence rate O(1/sqrt(k)). Sov Math Dokl 27:372–376

Pascanu R, Gülçehre Ç, Cho K, Bengio Y (2013a) How to construct deep recurrent neural networks. arXiv:1312.6026

Pascanu R, Mikolov T, Bengio Y (2012) Understanding the exploding gradient problem. Computing Research Repository (CoRR). arXiv:12115063

Pascanu R, Mikolov T, Bengio Y (2013b) On the difficulty of training recurrent neural networks. In: Proceedings of the 30th International conference on international conference on machine learning, JMLR.org, ICML'13, vol 28, pp III–1310–III–1318. http://dl.acm.org/citation.cfm?id=3042817.3043083

Pham V, Bluche T, Kermorvant C, Louradour J (2014a) Dropout improves recurrent neural networks for handwriting recognition. In: 2014 14th International Conference on Frontiers in Handwriting Recognition (ICFHR). IEEE, pp 285–290

Rumelhart DE, Hinton GE, Williams RJ (1985) Learning internal representations by error propagation, Technical report, DTIC Document

Scardapane S, Comminiello D, Hussain A, Uncini A (2017) Group sparse regularization for deep neural networks. Neurocomputing 241:81–89. https://doi.org/10.1016/j.neucom.2017.02.029

Scardapane S, Wang D (2017) Randomness in neural networks: an overview. Wiley Interdiscip Rev Data Min Knowl Discov 7(2):e1200. https://doi.org/10.1002/widm.1200

Schmidhuber J, Wierstra D, Gagliolo M, Gomez F (2007) Training recurrent networks by evolino. Neural Comput 19(3):757–779

Schoenholz SS, Gilmer J, Ganguli S, Sohl-Dickstein J (2016) Deep information propagation. arXiv:1611.01232

Schuster M, Paliwal KK (1997) Bidirectional recurrent neural networks. IEEE Trans Signal Process 45(11):2673–2681

Siegelmann HT, Sontag ED (1991) Turing computability with neural nets. Appl Math Lett 4(6):77–80

Srivastava RK, Greff K, Schmidhuber J (2015) Training very deep networks. In: Cortes C, Lawrence ND, Lee DD, Sugiyama M, Garnett R (eds) Advances in neural information processing systems, vol 28. Curran Associates Inc., pp 2377–2385

Sutskever I (2013) Training recurrent neural networks. PhD thesis, University of Toronto

Sutskever I, Hinton G (2010) Temporal-kernel recurrent neural networks. Neural Netw 23(2):239–243

Sutskever I, Martens J, Dahl GE, Hinton GE (2013) On the importance of initialization and momentum in deep learning. ICML 3(28):1139–1147

Sutskever I, Vinyals O, Le QV (2014) Sequence to sequence learning with neural networks. In: Advances in neural information processing systems, pp 3104–3112

Tieleman T, Hinton G (2012) Lecture 6.5-rmsprop: divide the gradient by a running average of its recent magnitude. COURSERA: Neural Netw Mach Learn 4:2

Williams RJ, Peng J (1990) An efficient gradient-based algorithm for on-line training of recurrent network trajectories. Neural Comput 2(4):490–501. https://doi.org/10.1162/neco.1990.2.4.490

Williams RJ, Zipser D (1995) Gradient-based learning algorithms for recurrent networks and their computational complexity. In: Backpropagation: theory, architectures, and applications, vol 1. pp 433–486

Williams RJ, Zipser D (1989) A learning algorithm for continually running fully recurrent neural networks. Neural Comput 1(2):270–280

Zhang S, Wu Y, Che T, Lin Z, Memisevic R, Salakhutdinov RR, Bengio Y (2016) Architectural complexity measures of recurrent neural networks. In: Lee DD, Sugiyama M, Luxburg UV, Guyon I, Garnett R (eds) Advances in Neural Information Processing Systems, vol 29. Curran Associates Inc., pp 1822–1830

Zilly JG, Srivastava RK, Koutník J, Schmidhuber J (2016) Recurrent highway networks. arXiv:1607:03474

Chapter 3
Recurrent Neural Network Architectures

Abstract In this chapter, we present three different recurrent neural network architectures that we employ for the prediction of real-valued time series. All the models
reviewed in this chapter can be trained through the previously discussed backpropagation through time procedure. First, we present the most basic version of recurrent
neural networks, called Elman recurrent neural network. Then, we introduce two
popular gated architectures, which are long short-term memory and the gated recurrent units. We discuss the main advantages of these more sophisticated architectures,
especially regarding their capability to process much longer dependencies in time
by maintaining an internal memory for longer periods. For each one of the reviewed
network, we provide the details and we show the equations for updating the internal
state and computing the output at each time step. Then, for each recurrent neural
network we also provide a quick overview of its main applications in previous works
in the context of real-valued time series forecasting.

Keywords Elman recurrent neural network · Gated architectures · Long short-term
memory · Gated recurrent unit · Time series prediction applications

3.1 Elman Recurrent Neural Network

The Elman Recurrent Neural Network (ERNN), also known as *Simple RNN* or *Vanilla
RNN*, is depicted in Fig. 2.1 and is usually considered to be the most basic version of
RNN. Most of the more complex RNN architectures, such as LSTM and GRU, can
be interpreted as a variation or as an extension of ERNNs.

ERNNs have been applied in many different contexts. In natural language processing applications, ERNN is demonstrated to be capable of learning grammar using
a training set of unannotated sentences to predict successive words in the sentence
Elman (1995), Ogata et al. (2007). Mori and Ogasawara (1993) studied ERNN performance in short-term load forecasting and proposed a learning method, called
"diffusion learning" (a sort of momentum-based gradient descent), to avoid local
minima during the optimization procedure. Cai et al. (2007) trained a ERNN with
a hybrid algorithm that combines particle swarm optimization and evolutionary

© The Author(s) 2017   23
F. M. Bianchi et al., *Recurrent Neural Networks for Short-Term Load Forecasting*,
SpringerBriefs in Computer Science, https://doi.org/10.1007/978-3-319-70338-1_3

computation to overcome the local minima issues of gradient-based methods. Furthermore, ERNNs have been employed by Cho (2003) in tourist arrival forecasting and by Mandal et al. (2006) to predict electric load time series. Due to the critical dependence of electric power usage on the day of the week or month of the year, a preprocessing step is performed to cluster similar days according to their load profile characteristics. Chitsaz et al. (2015) proposes a variant of ERNN called self-recurrent wavelet neural network, where the ordinary nonlinear activation functions of the hidden layer are replaced with wavelet functions. This leads to a sparser representation of the load profile, which is demonstrated to be helpful for tackling the forecast task through smaller and more easily trainable networks.

The layers in a RNN can be divided into *input layers*, *hidden layers*, and the *output layers* (see Fig. 2.1). While input and output layers are characterized by feedforward connections, the hidden layers contain recurrent ones. At each time step t, the input layer processes the component $\mathbf{x}[t] \in \mathbb{R}^{N_i}$ of a serial input $\mathbf{x}$. The time series $\mathbf{x}$ has length T and it can contain real values, discrete values, one-hot vectors, and so on. In the input layer, each component $\mathbf{x}[t]$ is summed with a bias vector $\mathbf{b}_i \in \mathbb{R}^{N_h}$ (N_h is the number of nodes in the hidden layer) and then is multiplied with the input weight matrix $\mathbf{W}_i^h \in \mathbb{R}^{N_i \times N_h}$. Analogously, the internal state of the network $\mathbf{h}[t-1] \in \mathbb{R}^{N_h}$ from the previous time interval is first summed with a bias vector $\mathbf{b}_h \in \mathbb{R}^{N_h}$ and then multiplied by the weight matrix $\mathbf{W}_h^h \in \mathbb{R}^{N_h \times N_h}$ of the recurrent connections. The transformed current input and past network state are then combined and processed by the neurons in the hidden layers, which apply a nonlinear transformation. The difference equations for the update of the internal state and the output of the network at a time step t are as follows:

$$
\begin{aligned}
\mathbf{h}[t] &= f\left(\mathbf{W}_i^h\left(\mathbf{x}[t] + \mathbf{b}_i\right) + \mathbf{W}_h^h\left(\mathbf{h}[t-1] + \mathbf{b}_h\right)\right), \\
\mathbf{y}[t] &= g\left(\mathbf{W}_h^o\left(\mathbf{h}[t] + \mathbf{b}_o\right)\right),
\end{aligned}
\tag{3.1}
$$

where $f(\cdot)$ is the activation function of the neurons, usually implemented by a sigmoid or by a hyperbolic tangent. The hidden state $\mathbf{h}[t]$ conveys the content of the memory of the network at time step t, is typically initialized with a vector of zeros and it depends on past inputs and network states. The output $\mathbf{y}[t] \in \mathbb{R}^{N_o}$ is computed through a transformation $g(\cdot)$, usually linear, on the matrix of the output weights $\mathbf{W}_h^o \in \mathbb{R}^{N_r \times N_o}$ applied to the sum of the current state $\mathbf{h}[t]$ and the bias vector $\mathbf{b}_o \in \mathbb{R}^{N_o}$. All the weight matrices and biases can be trained through gradient descent, according to the BPPT procedure. Unless differently specified, in the following to compact the notation we omit the bias terms by assuming $\mathbf{x} = [\mathbf{x}; 1]$, $\mathbf{h} = [\mathbf{h}; 1]$, $\mathbf{y} = [\mathbf{y}; 1]$ and by augmenting $\mathbf{W}_i^h$, $\mathbf{W}_h^h$, $\mathbf{W}_h^o$ with an additional column.

3.2 Long Short-Term Memory

The Long Short-Term Memory (LSTM) architecture was originally proposed by Hochreiter and Schmidhuber (1997) and is widely used nowadays due to its superior performance in accurately modeling both short- and long-term dependencies in data. LSTM tries to solve the vanishing gradient problem by not imposing any bias toward recent observations, but it keeps constant error flowing back through time. LSTM works essentially in the same way as the ERNN architecture, with the difference that it implements a more elaborated internal processing unit called *cell*.

LSTM has been employed in numerous sequence learning applications, especially in the field of natural language processing. Outstanding results with LSTM have been reached by Graves and Schmidhuber (2009) in unsegmented connected handwriting recognition, by Graves et al. (2013) in automatic speech recognition, by Eck and Schmidhuber (2002) in music composition and by Gers and Schmidhuber (2001) in grammar learning. Further successful results have been achieved in the context of image tagging, where LSTM have been paired with convolutional neural network to provide annotations on images automatically (Vinyals et al. 2017).

However, few works exist where LSTM has been applied to prediction of real-valued time series. Ma et al. (2015) evaluated the performances of several kinds of RNNs in short-term traffic speed prediction and compared them with other common methods like SVMs, ARIMA, and Kalman filters, finding that LSTM networks are nearly always the best approach. Pawlowski and Kurach (2015) utilized ensembles of LSTM and feedforward architectures to classify the danger from concentration level of methane in a coal mine, by predicting future concentration values. By following a hybrid approach, Felder et al. (2010) trains a LSTM network to output the parameter of a Gaussian mixture model that best fits a wind power temporal profile.

While an ERNN neuron implements a single nonlinearity $f(\cdot)$ (see Eq. 3.1), a LSTM cell is composed of 5 different nonlinear components, interacting with each other in a particular way. The internal state of a cell is modified by the LSTM only through linear interactions. This permits information to backpropagate smoothly across time, with a consequent enhancement of the memory capacity of the cell. LSTM protects and controls the information in the cell through three gates, which are implemented by a sigmoid and a pointwise multiplication. To control the behavior of each gate, a set of parameters are trained with gradient descent, in order to solve a target task.

Since its initial definition (Hochreiter and Schmidhuber 1997), several variants of the original LSTM unit have been proposed in the literature. In the following, we refer to the commonly used architecture proposed by Graves and Schmidhuber (2005). A schema of the LSTM cell is depicted in Fig. 3.1.

The difference equations that define the forward pass to update the cell state and to compute the output are listed below.

$$\text{forget gate}: \ \sigma_f[t] = \sigma\left(\mathbf{W}_f\mathbf{x}[t] + \mathbf{R}_f\mathbf{y}[t-1] + \mathbf{b}_f\right),$$
$$\text{candidate state}: \ \tilde{\mathbf{h}}[t] = g_1\left(\mathbf{W}_h\mathbf{x}[t] + \mathbf{R}_h\mathbf{y}[t-1] + \mathbf{b}_h\right),$$
$$\text{update gate}: \ \sigma_u[t] = \sigma\left(\mathbf{W}_u\mathbf{x}[t] + \mathbf{R}_u\mathbf{y}[t-1] + \mathbf{b}_u\right), \quad (3.2)$$
$$\text{cell state}: \ \mathbf{h}[t] = \sigma_u[t] \odot \tilde{\mathbf{h}}[t] + \sigma_f[t] \odot \mathbf{h}[t-1],$$
$$\text{output gate}: \ \sigma_o[t] = \sigma\left(\mathbf{W}_o\mathbf{x}[t] + \mathbf{R}_o\mathbf{y}[t-1] + \mathbf{b}_o\right),$$
$$\text{output}: \ \mathbf{y}[t] = \sigma_o[t] \odot g_2(\mathbf{h}[t]).$$

$\mathbf{x}[t]$ is the input vector at time t. $\mathbf{W}_f$, $\mathbf{W}_h$, $\mathbf{W}_u$, and $\mathbf{W}_o$ are rectangular weight matrices, that are applied to the input of the LSTM cell. $\mathbf{R}_f$, $\mathbf{R}_h$, $\mathbf{R}_u$, and $\mathbf{R}_o$ are square matrices that define the weights of the recurrent connections, while $\mathbf{b}_f$, $\mathbf{b}_h$, $\mathbf{b}_u$, and $\mathbf{b}_o$ are bias vectors. The function $\sigma(\cdot)$ is a sigmoid,[1] while $g_1(\cdot)$ and $g_2(\cdot)$ are pointwise nonlinear activation functions usually implemented as hyperbolic tangents that squash the values in $[-1, 1]$. Finally, $\odot$ is the entrywise multiplication between two vectors (Hadamard product).

Each gate in the cell has a specific and unique functionality. The *forget gate* σ_f decides what information should be discarded from the previous cell state $\mathbf{h}[t-1]$. The *input gate* σ_u operates on the previous state $\mathbf{h}[t-1]$, after having been modified by the forget gate, and it decides how much the new state $\mathbf{h}[t]$ should be updated with a new candidate $\tilde{\mathbf{h}}[t]$. To produce the output $\mathbf{y}[t]$, first the cell filters its current state with a nonlinearity $g_2(\cdot)$. Then, the *output gate* σ_o selects the part of the state to be returned as output. Each gate depends on the current external input $\mathbf{x}[t]$ and the previous cells output $\mathbf{y}[t-1]$.

As we can see from the Fig. 3.1 and from the forward step equations, when $\sigma_f = 1$ and $\sigma_u = 0$, the current state of a cell is transferred to the next time interval exactly as it is. By referring back to Eq. 2.10, it is possible to observe that in LSTM the issue of vanishing gradient does not occur, due to the absence of nonlinear transfer functions applied to the cell state. Since in this case the transfer function $f(\cdot)$ in Eq. 2.10 applied to the internal states is an identity function, the contribution from past states remains unchanged over time. However, in practice, the update and forget gates are never completely open or closed due to the functional form of the sigmoid, which saturates only for infinitely large values. As a result, even if long-term memory in LSTM is greatly enhanced with respect to ERNN architectures, the content of the cell cannot be kept completely unchanged over time.

3.3 Gated Recurrent Unit

The Gated Recurrent Unit (GRU) is another famous gated architecture, originally proposed by Cho et al. (2014), which adaptively captures dependencies at different time scales. In GRU, forget and input gates are combined into a single update gate,

---

[1] The logistic sigmoid is defined as $\sigma(x) = \frac{1}{1+e^{-x}}$.

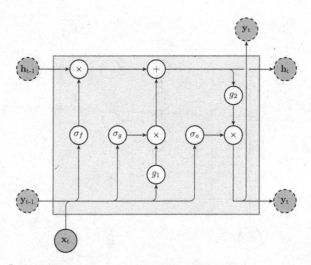

Fig. 3.1 Illustration of a cell in the LSTM architecture. Dark gray circles with a solid line are the variables whose content is exchanged with the input and output of the cell. Dark gray circles with a dashed line represent the internal state variables, whose content is exchanged between the cells of the hidden layer. Operators g_1 and g_2 are the nonlinear transformation usually implemented as a hyperbolic tangent. White circles with $+$ and $\times$ represent linear operations, while σ_f, σ_u, and σ_o are the sigmoids used in the forget, update, and output gates respectively

which adaptively controls how much each hidden unit can remember or forget. The internal state in GRU is always fully exposed in output, due to the lack of a control mechanism, like the output gate in LSTM.

GRUs were first tested by Cho et al. (2014) on a statistical machine translation task and reported mixed results. In an empirical comparison of GRU and LSTM, configured with the same amount of parameters, Chung et al. (2014) concluded that on some datasets GRU can outperform LSTM, both in terms of generalization capabilities and in terms of time required to reach convergence and to update parameters. In an extended experimental evaluation, (Zaremba 2015) employed GRU to (i) compute the digits of the sum or difference of two input numbers, (ii) predict the next character in a synthetic XML dataset and in the large words dataset Penn TreeBank, and (iii) predict polyphonic music. The results showed that the GRU outperformed the LSTM on nearly all tasks except language modeling when using a naive initialization. Bianchi et al. (2017) compared GRU with other recurrent networks on the prediction of superimposed oscillators. However, to the best of author's knowledge, at the moment there are no researches where the standard GRU architecture has been applied in STLF problems.

A schematic depiction of the GRU cell is reported in Fig. 3.2. GRU makes use of two gates. The first is the *update gate*, which controls how much the current content of the cell should be updated with the new candidate state. The second is the *reset gate* that, if closed (value near to 0), can effectively reset the memory of the cell and make the unit act as if the next processed input was the first in the sequence. The state equations of the GRU are the following:

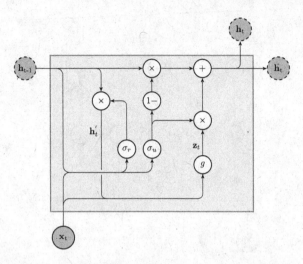

Fig. 3.2 Illustration of a recurrent unit in the GRU architecture. Dark gray circles with a solid line are the variables whose content is exchanged with the input and output of the network. Dark gray circles with a dashed line represent the internal state variables, whose content is exchanged within the cells of the hidden layer. The operator g is a nonlinear transformation, usually implemented as a hyperbolic tangent. White circles with "+", "−1", and "×" represent linear operations, while σ_r and σ_u are the sigmoids used in the reset and update gates respectively

$$
\begin{aligned}
\text{reset gate}: \quad & \mathbf{r}[t] = \sigma \left(\mathbf{W}_r \mathbf{h}[t-1] + \mathbf{R}_r \mathbf{x}[t] + \mathbf{b}_r \right), \\
\text{current state}: \quad & \mathbf{h}'[t] = \mathbf{h}[t-1] \odot \mathbf{r}[t], \\
\text{candidate state}: \quad & \mathbf{z}[t] = g \left(\mathbf{W}_z \mathbf{h}'[t-1] + \mathbf{R}_z \mathbf{x}[t] + \mathbf{b}_z \right), \\
\text{update gate}: \quad & \mathbf{u}[t] = \sigma \left(\mathbf{W}_u \mathbf{h}[t-1] + \mathbf{R}_u \mathbf{x}[t] + \mathbf{b}_u \right), \\
\text{new state}: \quad & \mathbf{h}[t] = (1 - \mathbf{u}[t]) \odot \mathbf{h}[t-1] + \mathbf{u}[t] \odot \mathbf{z}[t].
\end{aligned}
\tag{3.3}
$$

Here, $g(\cdot)$ is a nonlinear function usually implemented by a hyperbolic tangent.

In a GRU cell, the number of parameters is larger than in the ERNN unit, but smaller than in a LSTM cell. The parameters to be learned are the rectangular matrices $\mathbf{W}_r$, $\mathbf{W}_z$, $\mathbf{W}_u$, the square matrices $\mathbf{R}_r$, $\mathbf{R}_z$, $\mathbf{R}_u$, and the bias vectors $\mathbf{b}_r$, $\mathbf{b}_z$, $\mathbf{b}_u$.

References

Bianchi FM, Kampffmeyer M, Maiorino E, Jenssen R (2017) Temporal overdrive recurrent neural network. arXiv:170105159

Cai X, Zhang N, Venayagamoorthy GK, Wunsch DC (2007) Time series prediction with recurrent neural networks trained by a hybrid PSO-EA algorithm. Neurocomputing 70(13–15):2342–2353. https://doi.org/10.1016/j.neucom.2005.12.138

Chitsaz H, Shaker H, Zareipour H, Wood D, Amjady N (2015) Short-term electricity load forecasting of buildings in microgrids. Energy Build 99:50–60. https://doi.org/10.1016/j.enbuild.2015.04.011

Cho V (2003) A comparison of three different approaches to tourist arrival forecasting. Tour Manag 24(3):323–330. https://doi.org/10.1016/S0261-5177(02)00068-7

Cho K, Van Merriënboer B, Gulcehre C, Bahdanau D, Bougares F, Schwenk H, Bengio Y (2014) Learning phrase representations using RNN encoder-decoder for statistical machine translation. arXiv:14061078

Chung J, Gülçehre Ç, Cho K, Bengio Y (2014) Empirical evaluation of gated recurrent neural networks on sequence modeling. arXiv:1412.3555

Eck D, Schmidhuber J (2002) Finding temporal structure in music: Blues improvisation with LSTM recurrent networks. In: Proceedings of the 2002 12th IEEE workshop on neural networks for signal processing, 2002. IEEE, pp 747–756

Elman JL (1995) Language as a dynamical system. In: Mind as motion: explorations in the dynamics of cognition, pp 195–223

Felder M, Kaifel A, Graves A (2010) Wind power prediction using mixture density recurrent neural networks. Poster P0, vol 153, pp 1–7

Gers FA, Schmidhuber J (2001) Lstm recurrent networks learn simple context-free and context-sensitive languages. IEEE Trans Neural Netw 12(6):1333–1340

Graves A, Mohamed AR, Hinton G (2013) Speech recognition with deep recurrent neural networks. In: 2013 IEEE International Conference on Acoustics, Speech and Signal Processing (ICASSP). IEEE, pp 6645–6649

Graves A, Schmidhuber J (2005) Framewise phoneme classification with bidirectional LSTM and other neural network architectures. iJCNN 2005, Neural Netw 18(56):602–610. https://doi.org/10.1016/j.neunet.2005.06.042

Graves A, Schmidhuber J (2009) Offline handwriting recognition with multidimensional recurrent neural networks. In: Advances in neural information processing systems, pp 545–552

Hochreiter S, Schmidhuber J (1997) Long short-term memory. Neural Comput 9(8):1735–1780

Mandal P, Senjyu T, Urasaki N, Funabashi T (2006) A neural network based several-hour-ahead electric load forecasting using similar days approach. International J Electr Power Energy Syst 28(6):367–373. https://doi.org/10.1016/j.ijepes.2005.12.007

Ma X, Tao Z, Wang Y, Yu H, Wang Y (2015) Long short-term memory neural network for traffic speed prediction using remote microwave sensor data. Transp Res Part C Emerg Technol 54:187–197. https://doi.org/10.1016/j.trc.2015.03.014

Mori HMH, Ogasawara TOT (1993) A recurrent neural network for short-term load forecasting. In: 1993 proceedings of the second international forum on applications of neural networks to power systems, vol 31, pp 276–281. https://doi.org/10.1109/ANN.1993.264315

Ogata T, Murase M, Tani J, Komatani K, Okuno HG (2007) Two-way translation of compound sentences and arm motions by recurrent neural networks. In: IROS 2007, IEEE/RSJ International conference on intelligent robots and systems, 2007. IEEE, pp 1858–1863

Pawlowski K, Kurach K (2015) Detecting methane outbreaks from time series data with deep neural networks. In: Proceedings rough sets, fuzzy sets, data mining, and granular computing—15th international conference, RSFDGrC 2015, Tianjin, China, November 20–23, 2015, vol 9437, pp 475–484. https://doi.org/10.1007/978-3-319-25783-9_42

Vinyals O, Toshev A, Bengio S, Erhan D (2017) Show and tell: lessons learned from the 2015 MSCOCO image captioning challenge. IEEE Trans Pattern Anal Mach Intell 39(4):652–663. https://doi.org/10.1109/TPAMI.2016.2587640

Zaremba W (2015) An empirical exploration of recurrent network architectures. In: Proceedings of the 32nd international conference on machine learning, Lille, France

Chapter 4
Other Recurrent Neural Networks Models

Abstract In this chapter we review two additional types of Recurrent Neural Network, which present important differences with respect to the architectures described so far. More specifically, we introduce the nonlinear auto-regressive with eXogenous inputs (NARX) neural network and the Echo State Network. Both these networks have been largely employed in Short Term Load Forecast applications and they have been shown to be more effective than other methods based on statistical models. The main differences of NARX networks and Echo State Networks with respect to the other previously described models, are both in terms of their architecture and, in particular, in their training procedure. Indeed, both these architectures are designed in such a way that Back Propagation Through Time is not necessary. Specifically, in NARX the network output is replaced by the expected ground truth and this allows to train the network like a feedforward architecture. On the other hand, in a Echo State Network only the outermost linear layer is trained, usually by means of ridge regression. Due to these fundamental differences, some of the properties and training approaches discussed in the previous sections do not hold for the NARX and Echo State Network models and we reserved a separate chapter to review these models.

Keywords nonlinear auto-regressive eXogenous inputs neural network · feedforward neural networks · Echo state network · Reservoir computing · Randomized neural network · Linear regression

4.1 NARX Network

NARX networks are recurrent dynamic architectures with several hidden layers and they are inspired by discrete-time nonlinear models called nonlinear auto-regressive with eXogenous inputs (Leontaritis and Billings 1985). Differently from other RNNs, the recurrence in the NARX network is given only by the feedback on the output, rather than from the whole internal state.

NARX networks have been employed in many different applicative contexts, to forecast future values of the input signal (Diaconescu 2008; Lin et al. 1997). Menezes and Barreto (2008) showed that NARX networks perform better on predictions

© The Author(s) 2017

F. M. Bianchi et al., *Recurrent Neural Networks for Short-Term Load Forecasting*, SpringerBriefs in Computer Science, https://doi.org/10.1007/978-3-319-70338-1_4

involving long-term dependencies. (Xie et al., 2009) used NARX in conjunction with an input embedded according to Takens method, to predict highly nonlinear time series. NARX are also employed as a nonlinear filter, whose target output is trained by using the noise-free version of the input signal (Napoli and Piroddi 2010). NARX networks have also been adopted by Plett (2003) in a gray-box approach for nonlinear system identification.

A NARX network can be implemented with a multilayer perceptron (MLP), where the next value of the output signal $\mathbf{y}[t] \in \mathbb{R}^{N_y}$ is regressed on d_y previous values of the output signal and on d_x previous values of an independent, exogenous input signal $\mathbf{x}[t] \in \mathbb{R}^{N_x}$ (Billings 2013). The output equation reads

$$\mathbf{y}[t] = \phi\left(\mathbf{x}[t - d_x], \ldots, \mathbf{x}[t - 1], \mathbf{x}[t], \mathbf{y}[t - d_y], \ldots, \mathbf{y}[t - 1], \Theta\right), \qquad (4.1)$$

where $\phi(\cdot)$ is the nonlinear mapping function performed by the MLP, Θ are the trainable network parameters, d_x and d_y are the input and the output time delays. Even if the numbers of delays d_x and d_y is a finite (often small) number, it has been proven that NARX networks are at least as powerful as Turing machines, and thus they are universal computation devices (Siegelmann et al. 1997).

The input $\mathbf{i}[t]$ of the NARX network has $d_x N_x + d_y N_y$ components, which correspond to a set of two Tapped-Delay Lines (TDLs), and it reads

$$\mathbf{i}[t] = \begin{bmatrix} (\mathbf{x}[t - d_x], \ldots, \mathbf{x}[t - 1])^T \\ (\mathbf{y}[t - d_y], \ldots, \mathbf{y}[t - 1])^T \end{bmatrix}^T. \qquad (4.2)$$

The structure of a MLP network consists of a set of source nodes forming the input layer, $N_l \geq 1$ layers of hidden nodes, and an output layer of nodes. The output of the network is governed by the following difference equations

$$\mathbf{h}_1[t] = f\left(\mathbf{i}[t], \theta_i\right), \qquad (4.3)$$

$$\mathbf{h}_l[t] = f\left(\mathbf{h}_{l-1}[t - 1], \theta_{h_l}\right), \qquad (4.4)$$

$$\mathbf{y}[t] = g\left(\mathbf{h}_{N_l}[t - 1], \theta_o\right), \qquad (4.5)$$

where $\mathbf{h}_l[t] \in \mathbb{R}^{N_{h_l}}$ is the output of the lth hidden layer at time t, $g(\cdot)$ is a linear function and $f(\cdot)$ is the transfer function of the neuron, usually implemented as a sigmoid or *tanh* function.

The weights of the neurons connections are defined by the parameters

$$\Theta = \left\{\theta_i, \theta_o, \theta_{h_1}, \ldots, \theta_{h_{N_l}}\right\}.$$

In particular, $\theta_i = \left\{\mathbf{W}_i^{h_1} \in \mathbb{R}^{d_x N_x + d_y N_y \times N_{h_1}}, \mathbf{b}_{h_1} \in \mathbb{R}^{N_{h_1}}\right\}$ are the parameters that determine the weights in the input layer, $\theta_o = \left\{\mathbf{W}_{h_{N_l}}^o \in \mathbb{R}^{N_{N_l} \times N_y}, \mathbf{b}_o \in \mathbb{R}^{N_y}\right\}$ are the parameters of the output layer and $\theta_{h_l} = \left\{ \mathbf{W}^{h_l}_{h_{l-1}} \in \mathbb{R}^{N_{h_{l-1}} \times N_{h_l}}, \mathbf{b}_{h_l} \in \mathbb{R}^{N_{h_l}} \right\}$ are the parameters of the lth hidden layer. A schematic depiction of a NARX network is reported in Fig. 4.1.

Due to the architecture of the network, it is possible to exploit a particular strategy to learn the parameters Θ. Specifically, during the training phase the time series relative to the desired output $\mathbf{y}^*$ is fed into the network along with the input time series $\mathbf{x}$. At this stage, the output feedback is disconnected and the network has a purely feedforward architecture, whose parameters can be trained with one of the several, well-established standard backpropagation techniques. Notice that this operation is not possible in other recurrent networks such as ERNN, since the state of the hidden layer depends on the previous hidden state, whose ideal value is not retrievable from the training set. Once the training stage is over, the teacher signal of the desired output is disconnected and is replaced with the feedback of the predicted output $\mathbf{y}$ computed by the network. The procedure is depicted in Fig. 4.2.

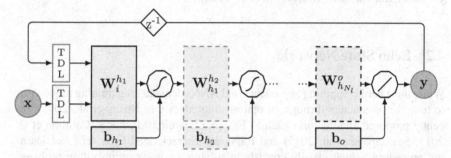

Fig. 4.1 Architecture of the NARX network. Circles represent input $\mathbf{x}$ and output $\mathbf{y}$, respectively. The two TDL blocks are the tapped-delay lines. The solid squares $\mathbf{W}^{h_1}_i$, $\mathbf{W}^o_{h_{N_l}}$, $\mathbf{b}_i$, and $\mathbf{b}_o$ are the weight matrices and the bias relative to the input and the output respectively. The dashed squares are the weight matrices and the biases relative to the N_l hidden layers – in the figure, we report $\mathbf{W}^{h_2}_{h_1}$ and $\mathbf{b}_{h_2}$, relative to the first hidden layer. The polygon with the sigmoid symbol represents the nonlinear transfer function of the neurons and the one with the oblique line is a linear operation. Finally, z^{-1} is the backshift/lag operator

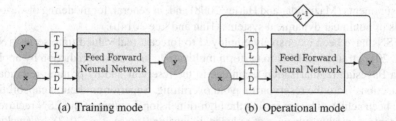

(a) Training mode   (b) Operational mode

Fig. 4.2 During the training, the desired input $\mathbf{y}^*$ is fed directly to the network. Once the network parameters have been optimized, the teacher signal is removed and the output $\mathbf{y}$ produced by the network is connected to the input with a feedback loop

Similar to what discussed in Sect. 2.2 for the previous RNN architectures, the loss function employed in the gradient descent is defined as

$$L(\mathbf{x}, \mathbf{y}^*; \Theta) = \mathrm{MSE}(\mathbf{y}, \mathbf{y}^*) + \lambda_2 \|\Theta\|_2, \tag{4.6}$$

where MSE is the error term defined in Eq. 2.2 and λ_2 is the hyperparameter that weights the importance of the L_2 regularization term in the loss function. Due to the initial transient phase of the network, when the estimated output $\mathbf{y}$ is initially fed back as network input, the first initial outputs are discarded.

Even if it reduces to a feedforward network in the training phase, NARX network is not immune to the problem of vanishing and exploding gradients. This can be seen by looking at the Jacobian $\mathbf{J_h}(t, n)$ of the state-space map at time t expanded for n time step. In order to guarantee network stability, the Jacobian must have all of its eigenvalues inside the unit circle at each time step. However, this results in $\lim_{n \to \infty} \mathbf{J_h}(t, n) = 0$, which implies that NARX networks suffer from vanishing gradients, like the other RNNs (Lin et al. 1996).

4.2   Echo State Network

While most hard computing approaches and ANNs demand long training procedures to tune the parameters through an optimization algorithm Huang et al. (2005), recently proposed architectures such as Extreme Learning Machines (Cambria et al. 2013; Scardapane et al. 2015) and ESNs are characterized by a very fast learning procedure, which usually consists in solving a convex optimization problem. ESNs, along with Liquid State Machines (Maass et al. 2002), belong to the class of computational dynamical systems implemented according to the so-called *reservoir computing* framework (Lukoševičius and Jaeger 2009).

ESN have been applied in a variety of different contexts, such as static classification (Alexandre et al. 2009), speech recognition (Skowronski and Harris. 2007), intrusion detection (Hai-yan et al. 2005), adaptive control (Han and Lee 2014a), detrending of nonstationary time series (Maiorino et al. 2017), harmonic distortion measurements (Mazumdar and Harley 2008) and, in general, for modeling of various kinds of nonlinear dynamical systems (Han and Lee 2014b).

ESNs have been extensively employed to forecast real valued time series. (Niu et al. 2012) trained an ESN to perform multivariate time series prediction by applying a Bayesian regularization technique to the reservoir and by pruning redundant connections from the reservoir to avoid overfitting. Superior prediction capabilities have been achieved by projecting the high-dimensional output of the ESN recurrent layer into a suitable subspace of reduced dimension (Løkse et al. 2017), by stacking multiple reservoirs to develop and enhance hierarchical dynamics (Gallicchio et al. 2017), or by designing simpler fixed reservoir topologies (Rodan and Tiňo 2012). An important context of application with real valued time series is the prediction

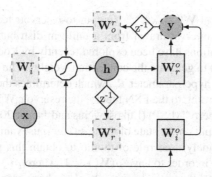

Fig. 4.3 Schematic depiction of the ESN architecture. The circles represent input **x**, state **h**, and output **y**, respectively. Solid squares $\mathbf{W}_r^o$ and $\mathbf{W}_i^o$, are the trainable matrices of the readout, while dashed squares, $\mathbf{W}_r^r$, $\mathbf{W}_o^r$, and $\mathbf{W}_i^r$, are randomly initialized matrices. The polygon represents the nonlinear transformation performed by neurons and z^{-1} is the unit delay operator

of telephonic or electricity load, usually performed 1 h and a 24 h ahead (Deihimi and Showkati 2012; Deihimi et al. 2013; Bianchi et al 2015; Varshney and Verma 2014; Bianchi et al. 2015b). (Deihimi et al. 2013; Peng et al. 2014) decomposed the time series in wavelet components, which are predicted separately using distinct ESN and ARIMA models, whose outputs are combined to produce the final result. Important results have been achieved in the prediction of chaotic time series by Li et al. (2012a). They proposed an alternative to the Bayesian regression for estimating the regularization parameter and a Laplacian likelihood function, more robust to noise and outliers than a Gaussian likelihood. Jaeger and Haas (2004) applied an ESN-based predictor on both benchmark and real dataset, highlighting the capability of these networks to learn amazingly accurate models to forecast a chaotic process from almost noise-free training data.

An ESN consists of a large, sparsely connected, untrained recurrent layer of nonlinear units and a linear, memory-less read-out layer, which is trained according to the task that the ESN is demanded to solve. A visual representation of an ESN is shown in Fig. 4.3

The difference equations describing the ESN state-update and output are, respectively, defined as follows:

$$\mathbf{h}[t] = f(\mathbf{W}_r^r \mathbf{h}[t-1] + \mathbf{W}_i^r \mathbf{x}[t] + \mathbf{W}_o^r \mathbf{y}[t-1] + \epsilon), \qquad (4.7)$$

$$\mathbf{y}[t] = g(\mathbf{W}_i^o \mathbf{x}[t] + \mathbf{W}_r^o \mathbf{h}[t]), \qquad (4.8)$$

where ϵ is a small noise term. The reservoir contains N_h neurons whose transfer/activation function $f(\cdot)$ is typically implemented by a hyperbolic tangent. The readout instead, is implemented usually by a linear function $g(\cdot)$. At time instant t, the network is driven by the input signal $\mathbf{x}[t] \in \mathbb{R}^{N_i}$ and produces the output $\mathbf{y}[k] \in \mathbb{R}^{N_o}$, N_i and N_o being the dimensionality of input and output, respectively. The vector $\mathbf{h}[t]$ has N_h components and it describes the ESN (instantaneous) state. The weight matrices $\mathbf{W}_r^r \in \mathbb{R}^{N_r \times N_r}$ (reservoir connections), $\mathbf{W}_i^r \in \mathbb{R}^{N_i \times N_r}$

(input-to-reservoir), and $\mathbf{W}_o^r \in \mathbb{R}^{N_o \times N_r}$ (output-to-reservoir feedback) contain real values in the $[-1, 1]$ interval drawn from a uniform distribution and are left un-trained. Alternative options have been explored recently by Rodan and Tiňo (2011), Appeltant et al. (2011) to generate the connection weights. The sparsity of the reservoir is controlled by a hyperparameter R_c, which determines the number of nonzero elements in $\mathbf{W}_r^r$. According to the ESN theory, the reservoir $\mathbf{W}_r^r$ must satisfy the so-called "echo state property" (ESP) (Lukoševičius and Jaeger 2009). This means that the effect of a given input on the state of the reservoir must vanish in a finite number of time-instants. A widely used rule-of-thumb to obtain this property suggests to rescale the matrix $\mathbf{W}_r^r$ in order to have $\rho(\mathbf{W}_r^r) < 1$, where $\rho(\cdot)$ denotes the spectral radius. However, several theoretical approaches have been proposed in the literature to tune ρ more accurately, depending on the problem at hand (Boedecker et al 2012; Bianchi et al. 2016a; Verstraeten and Schrauwen 2009; Bianchi et al. 2017).

On the other hand, the weight matrices $\mathbf{W}_i^o$ and $\mathbf{W}_r^o$ are optimized for the target task. To determine them, let us consider the training sequence of T_{tr} desired input-outputs pairs given by:

$$(\mathbf{x}[1], y^*[1]) \ldots, (\mathbf{x}[T_{\text{tr}}], y[T_{\text{tr}}]),\tag{4.9}$$

where T_{tr} is the length of the training sequence. In the initial phase of training, called *state harvesting*, the inputs are fed to the reservoir, producing a sequence of internal states $\mathbf{h}[1], \ldots, \mathbf{h}[T_{\text{tr}}]$, as defined in Eq. (4.7). The states are stacked in a matrix $\mathbf{S} \in \mathbb{R}^{T_{\text{tr}} \times N_i + N_r}$ and the desired outputs in a vector $\mathbf{y}^* \in \mathbb{R}^{T_{\text{tr}}}$:

$$\mathbf{S} = \begin{bmatrix} \mathbf{x}^T[1], & \mathbf{h}^T[1] \\ \vdots & \\ \mathbf{x}^T[T_{\text{tr}}], & \mathbf{h}^T[T_{\text{tr}}] \end{bmatrix},\tag{4.10}$$

$$\mathbf{y}^* = \begin{bmatrix} \mathbf{y}^*[1] \\ \vdots \\ \mathbf{y}^*[T_{\text{tr}}] \end{bmatrix}.\tag{4.11}$$

The initial rows in $\mathbf{S}$ (and $\mathbf{y}^*$) are discarded, since they refer to a transient phase in the ESN's behavior.

The training of the readout consists in learning the weights in $\mathbf{W}_i^o$ and $\mathbf{W}_r^o$ so that the output of the ESN matches the desired output $\mathbf{y}^*$. This procedure is termed *teacher forcing* and can be accomplished by solving a convex optimization problem, for which several closed form solution exist in the literature. The standard approach, originally proposed by Jaeger (2001), consists in applying a least-square regression, defined by the following regularized least-square problem:

$$W_{\text{ls}}^* = \underset{W \in \mathbb{R}^{N_i + N_h}}{\arg\min} \frac{1}{2} \|SW - \mathbf{y}^*\|^2 + \frac{\lambda_2}{2} \|W\|_2^2,\tag{4.12}$$

where $W = \begin{bmatrix} \mathbf{W}_i^o , \mathbf{W}_r^o \end{bmatrix}^T$ and $\lambda_2 \in \mathbb{R}^+$ is the L_2 regularization factor.

A solution to problem (4.12) can be expressed in closed form as

$$W_{1s}^* = \left(\mathbf{S}^T \mathbf{S} + \lambda_2 \mathbf{I}\right)^{-1} \mathbf{S}^T \mathbf{y}^*, \tag{4.13}$$

which can be solved by computing the Moore-Penrose pseudo-inverse. Whenever $N_h + N_i > T_{\mathrm{tr}}$, Eq. (4.13) can be computed more efficiently by rewriting it as

$$W_{1s}^* = \mathbf{S}^T \left(\mathbf{S}\mathbf{S}^T + \lambda_2 \mathbf{I}\right)^{-1} \mathbf{y}^*. \tag{4.14}$$

References

Alexandre LA, Embrechts MJ, Linton J (2009) Benchmarking reservoir computing on time-independent classification tasks. In: 2009 International Joint Conference on Neural Networks, IJCNN 2009. IEEE, pp 89–93

Appeltant L, Soriano MC, Van der Sande G, Danckaert J, Massar S, Dambre J, Schrauwen B, Mirasso CR, Fischer I (2011) Information processing using a single dynamical node as complex system. Nature Commun 2:468. https://doi.org/10.1038/ncomms1476

Bianchi FM, De Santis E, Rizzi A, Sadeghian A (2015) Short-term electric load forecasting using echo state networks and PCA decomposition. IEEE Access 3:1931–1943. https://doi.org/10.1109/ACCESS.2015.2485943

Bianchi FM, Scardapane S, Uncini A, Rizzi A, Sadeghian A (2015) Prediction of telephone calls load using echo state network with exogenous variables. Neural Netw 71:204–213. https://doi.org/10.1016/j.neunet.2015.08.010

Bianchi FM, Livi L, Alippi C (2016a) Investigating echo-state networks dynamics by means of recurrence analysis. IEEE Trans Neural Netw Learn Syst 99:1–13. https://doi.org/10.1109/TNNLS.2016.2630802

Billings SA (2013) Nonlinear system identification: NARMAX methods in the time, frequency, and spatio-temporal domains. Wiley

Boedecker J, Obst O, Lizier JT, Mayer NM, Asada M (2012) Information processing in echo state networks at the edge of chaos. Theory Biosci 131(3):205–213

Cambria E, Huang GB, Kasun LLC, Zhou H, Vong CM, Lin J, Yin J, Cai Z, Liu Q, Li K et al (2013) Extreme learning machines [trends & controversies]. IEEE Intell Syst 28(6):30–59

Deihimi A, Showkati H (2012) Application of echo state networks in short-term electric load forecasting. Energy 39(1):327–340

Deihimi A, Orang O, Showkati H (2013) Short-term electric load and temperature forecasting using wavelet echo state networks with neural reconstruction. Energy 57:382–401

Diaconescu E (2008) The use of narx neural networks to predict chaotic time series. Wseas Trans Comput Res 3(3):182–191

Filippo MB, Livi L, Alippi C, Jenssen R (2017) Multiplex visibility graphs to investigate recurrent neural network dynamics. Sci Rep 7(44):037. https://doi.org/10.1038/srep4403710.1038/srep44037

Gallicchio C, Micheli A, Pedrelli L (2017) Deep reservoir computing: a critical experimental analysis. Neurocomputing. https://doi.org/10.1016/j.neucom.2016.12.089

Hai-yan D, Wen-jiang P, Zhen-ya H (2005) A multiple objective optimization based echo state network tree and application to intrusion detection. In: 2005 Proceedings of 2005 IEEE international workshop on VLSI design and video technology, pp 443–446. doi:https://doi.org/10.1109/IWVDVT.2005.1504645

Han S, Lee J (2014a) Fuzzy echo state neural networks and funnel dynamic surface control for prescribed performance of a nonlinear dynamic system. IEEE Trans Ind Electron 61(2):1099–1112. https://doi.org/10.1109/TIE.2013.2253072

Han SI, Lee JM (2014b) Fuzzy echo state neural networks and funnel dynamic surface control for prescribed performance of a nonlinear dynamic system. IEEE Trans Ind Electron 61(2):1099–1112

Huang CM, Huang CJ, Wang ML (2005) A particle swarm optimization to identifying the armax model for short-term load forecasting. IEEE Trans Power Syst 20(2):1126–1133

Jaeger H (2001) The "echo state" approach to analysing and training recurrent neural networks-with an erratum note. Technical Report 148:34, German National Research Center for Information Technology GMD, Bonn, Germany

Jaeger H, Haas H (2004) Harnessing nonlinearity: Predicting chaotic systems and saving energy in wireless communication. Science 304(5667):78–80. https://doi.org/10.1126/science.1091277

Leontaritis I, Billings SA (1985) Input-output parametric models for non-linear systems part i: deterministic non-linear systems. Int J Control 41(2):303–328

Li D, Han M, Wang J (2012a) Chaotic time series prediction based on a novel robust echo state network. IEEE Trans Neural Netw Learn Syst 23(5):787–799

Lin T, Horne BG, Tiňo P, Giles CL (1996) Learning long-term dependencies in narx recurrent neural networks. IEEE Trans Neural Netw 7(6):1329–1338

Lin TN, Giles CL, Horne BG, Kung SY (1997) A delay damage model selection algorithm for narx neural networks. IEEE Trans Signal Process 45(11):2719–2730

Løkse S, Bianchi FM, Jenssen R (2017) Training echo state networks with regularization through dimensionality reduction. Cogn Comput 1–15. doi:https://doi.org/10.1007/s12559-017-9450-z

Lukoševičius M, Jaeger H (2009) Reservoir computing approaches to recurrent neural network training. Comput Sci Rev 3(3):127–149. https://doi.org/10.1016/j.cosrev.2009.03.005

Maass W, Natschläger T, Markram H (2002) Real-time computing without stable states: a new framework for neural computation based on perturbations. Neural Comput 14(11):2531–2560. https://doi.org/10.1162/089976602760407955

Maiorino E, Bianchi F, Livi L, Rizzi A, Sadeghian A (2017) Data-driven detrending of nonstationary fractal time series with echo state networks. Inf Sci 382–383:359–373. https://doi.org/10.1016/j.ins.2016.12.015

Mazumdar J, Harley R (2008) Utilization of echo state networks for differentiating source and nonlinear load harmonics in the utility network. IEEE Trans Power Electron 23(6):2738–2745. https://doi.org/10.1109/TPEL.2008.2005097

Menezes JMP, Barreto GA (2008) Long-term time series prediction with the narx network: an empirical evaluation. Neurocomputing 71(16):3335–3343

Napoli R, Piroddi L (2010) Nonlinear active noise control with narx models. IEEE Trans Audio, Speech, and Lang Process 18(2):286–295

Niu D, Ji L, Xing M, Wang J (2012) Multi-variable echo state network optimized by Bayesian regulation for daily peak load forecasting. J Netw 7(11):1790–1795

Peng Y, Lei M, Li JB, Peng XY (2014) A novel hybridization of echo state networks and multiplicative seasonal ARIMA model for mobile communication traffic series forecasting. Neural Comput Appl 24(3–4):883–890

Plett GL (2003) Adaptive inverse control of linear and nonlinear systems using dynamic neural networks. IEEE Trans Neural Netw 14(2):360–376

Rodan A, Tiňo P (2011) Minimum complexity echo state network. IEEE Trans Neural Netw 22(1):131–144. https://doi.org/10.1109/TNN.2010.2089641

Rodan A, Tiňo P (2012) Simple deterministically constructed cycle reservoirs with regular jumps. Neural Comput 24(7):1822–1852. https://doi.org/10.1162/NECO_a_00297

Scardapane S, Comminiello D, Scarpiniti M, Uncini A (2015) Online sequential extreme learning machine with kernels. IEEE Trans Neural Netw Learn Syst 26(9):2214–2220. https://doi.org/10.1109/TNNLS.2014.2382094

Siegelmann HT, Horne BG, Giles CL (1997) Computational capabilities of recurrent narx neural networks. IEEE Trans Syst Man Cybern Part B: Cybern 27(2):208–215

Skowronski MD, Harris JG (2007) Automatic speech recognition using a predictive echo state network classifier. Neural Netw 20(3):414–423

Varshney S, Verma T (2014) Half hourly electricity load prediction using echo state network. Int J Sci Res 3(6):885–888

Verstraeten D, Schrauwen B (2009) On the quantification of dynamics in reservoir computing. In: Alippi C, Polycarpou M, Panayiotou C, Ellinas G (eds) Artificial neural networks—ICANN 2009, vol 5768. Springer, Berlin, pp 985–994. doi:https://doi.org/10.1007/978-3-642-04274-4_101

Xie H, Tang H, Liao YH (2009) Time series prediction based on narx neural networks: An advanced approach. In: 2009 International conference on machine learning and cybernetics, vol 3, pp 1275–1279. doi:https://doi.org/10.1109/ICMLC.2009.5212326

Chapter 5
Synthetic Time Series

Abstract In this chapter, we describe three different synthetic datasets that we considered to evaluate the performance of the reviewed recurrent neural network architectures in a controlled environment. The generative models of the synthetic time series are the Mackey–Glass system, NARMA, and multiple superimposed oscillators.Those are benchmark tasks commonly considered in the literature to evaluate the performance of a predictive model. The three forecasting exercises that we study have varying levels of difficulty, given by the nature of the signal and the complexity of the task to be solved by the RNN.

Keywords Synthetic time series · Benchmark prediction tasks · Mackey–Glass system · Nonlinear auto-regressive moving average task · Multiple superimposed oscillators

In order to obtain a prediction problem that is not too simple, it is reasonable to select as forecast horizon a time interval t_f that guarantees the measurements in the time series to become decorrelated. Hence, we consider the first zero of the autocorrelation function of the time series. Alternatively, the first minimum of the average mutual information Fraser and Swinney (1986) or of the correlation sum Liebert and Schuster (1989) could be chosen to select a t_f where the signal shows a more general form of independence. All the time series introduced in the following consist of 15.000 time steps. We use the first 60% of the time series as training set, to learn the parameters of the RNN models. The next 20% of the data are used as validation set and the prediction accuracy achieved by the RNNs on this second dataset is used to tune the hyperparameters of the models. The final model performance is evaluated on a test set, corresponding to the last 20% of the values in the time series.

Mackey–Glass Time Series

The Mackey-Glass (MG) system is commonly used as benchmark for prediction of chaotic time series. The input signal is generated from the MG time delay differential system, described by the following equation:

$$\frac{dx}{dt} = \frac{\alpha x(t - \tau_{MG})}{1 + x(t - \tau_{MG})^{10}} - \beta x(t). \tag{5.1}$$

© The Author(s) 2017   41
F. M. Bianchi et al., *Recurrent Neural Networks for Short-Term Load Forecasting*,
SpringerBriefs in Computer Science, https://doi.org/10.1007/978-3-319-70338-1_5

For this prediction task, we set $\tau_{MG} = 17$, $\alpha = 0.2$, $\beta = 0.1$, initial condition $x(0) = 1.2, 0.1$ as integration step for (5.1) and the forecast horizon $t_f = 12$.

NARMA Signal

The nonlinear auto-regressive moving average (NARMA) task, originally proposed by Jaeger (2002a), consists in modeling the output of the following r-order system:

$$y(t + 1) = 0.3y(t) + 0.05y(t) \left[\sum_{i=0}^{r} y(t - i) \right] + 1.5x(t - r)x(t) + 0.1. \quad (5.2)$$

The input to the system $x(t)$ is uniform random noise in $[0, 1]$, and the model is trained to reproduce $y(t + 1)$. The NARMA task is known to require a memory of at least r past time steps, since the output is determined by input and outputs from the last r time steps. For this prediction task, we set $r = 10$ and the forecast step $t_f = 1$ in our experiments.

Multiple Superimposed Oscillator

The prediction of a sinusoidal signal is a relatively simple task, which demands a minimum amount of memory to determine the next network output. However, superimposed sine waves with incommensurable frequencies are extremely difficult to predict, since the periodicity of the resulting signal is extremely long. The time series we consider is the multiple superimposed oscillator (MSO) introduced by Jaeger and Haas (2004), and it is defined as

$$y(t) = sin(0.2t) + sin(0.311t) + sin(0.42t) + sin(0.51t). \quad (5.3)$$

This academic, yet important task, is particularly useful to test the memory capacity of a recurrent neural network and has been studied in detail by Xue et al. (2007) in a dedicated work. Indeed, to accurately predict the unseen values of the time series, the network requires a large amount of memory to simultaneously implement multiple decoupled internal dynamics Wierstra et al. (2005). For this last prediction task, we chose a forecast step $t_f = 10$.

References

Fraser AM, Swinney HL (1986) Independent coordinates for strange attractors from mutual information. Phys Rev A 33(2):1134

Liebert W, Schuster H (1989) Proper choice of the time delay for the analysis of chaotic time series. Phys Lett A 142(2–3):107–111

Jaeger H (2002a) Adaptive nonlinear system identification with echo state networks. In: Advances in neural information processing systems, pp 593–600

Jaeger H, Haas H (2004) Harnessing nonlinearity: Predicting chaotic systems and saving energy in wireless communication. Science 304(5667):78–80. https://doi.org/10.1126/science.1091277

Xue Y, Yang L, Haykin S (2007) Decoupled echo state networks with lateral inhibition. Neural Netw 20(3):365 – 376. http://dx.doi.org/10.1016/j.neunet.2007.04.014 (echo State Networks and Liquid State Machines)

Wierstra D, Gomez FJ, Schmidhuber J (2005) Modeling systems with internal state using evolino. In: Proceedings of the 7th annual conference on Genetic and evolutionary computation. pp 1795–1802. ACM

Chapter 6
Real-World Load Time Series

Abstract In this chapter, we consider three different real-world datasets, which contain real-valued time series of measurements of electricity and telephonic activity load. For each dataset, we set up a short-term load forecast problem of 24 hours ahead prediction. Two of the datasets under analysis include time series of measurements of exogenous variables, which are used to provide additional context to the network and thus to improve the accuracy of the prediction. For each dataset, we perform an analysis to study the nature of the time series, in terms of its correlation properties, seasonal patterns, correlation with the exogenous time series, and nature of the variance. According to the result of our analysis, we select a suitable preprocessing strategy before feeding the data into the recurrent neural networks. As shown in the following, the forecast accuracy in a prediction problem can be considerably improved by proper preprocessing of data (Zhang and Qi 2005).

Keywords Electricity load dataset · Telephonic activity load dataset · Call data records dataset · Seasonal pattern analysis · Correlation and autocorrelation analysis · Variance analysis · Time series data pre-preprocessing

6.1 Orange Dataset—Telephonic Activity Load

The first real-world dataset that we analyze is relative to the load of phone calls registered over a mobile network. Data come from the Orange telephone dataset (Orange 2013), published in the Data for Development (D4D) challenge (Blondel et al. 2012). D4D is a collection of call data records, containing anonymized events of Orange's mobile phone users in Ivory Coast, in a period spanning from December 1, 2011 to April 28, 2012. More detailed information on the data are available in Ref. Bianchi et al. (2016). The time series we consider are relative to antenna-to-antenna traffic. In particular, we selected a specific antenna, retrieved all the records in the dataset relative to the telephone activity issued each hour in the area covered by the antenna and generated six time series:

- ts1: number of incoming calls in the area covered by the antenna;
- ts2: volume in minutes of the incoming calls in the area covered by the antenna;

F. M. Bianchi et al., *Recurrent Neural Networks for Short-Term Load Forecasting*,
SpringerBriefs in Computer Science, https://doi.org/10.1007/978-3-319-70338-1_6

Fig. 6.1 In **a**, the load
profile of ts1, the incoming
calls volume, for 300 time
intervals (hours). In **b**, the
autocorrelation functions of
the time series ts1 before
(gray line) and after (black
line) a seasonal
differentiation. The original
time series shows a strong
seasonal pattern at lag 24,
while after seasonal
differencing, the time series
does not show any strong
correlation or trend

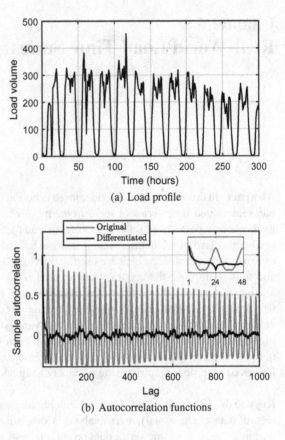

(a) Load profile

(b) Autocorrelation functions

- ts3: number of outgoing calls in the area covered by the antenna;
- ts4: volume in minutes of the outgoing calls in the area covered by the antenna;
- ts5: hour when the telephonic activity is registered; and
- ts6: day when the telephonic activity is registered.

In this work, we focus on predicting the volume (in minutes) of the incoming calls
in ts1 of the next day. Due to the hourly resolution of the data, the STFL problem
consists of a 24-step-ahead prediction. The profile of ts1 for 300 hours is depicted in
Fig. 6.1a. The remaining time series are treated as exogenous variables and, according
to a common practice in time series forecasting (Franses 1991), they are fed into
the network to provide the model with additional information for improving the
prediction of the target time series. Each time series contains 3336 measurements,
hourly sampled. We used the first 70% as training set, the successive 15% as validation
set, and the remaining 15% as test set. The accuracy of each RNN model is evaluated
on this last set.

In each time series, there is a (small) fraction of missing values. In fact, if in a given
hour no activities are registered in the area covered by the considered antenna, the
relative entries do not appear in the database. As we require the target time series and

the exogenous ones to have same lengths and to contain a value in each time interval, we inserted an entry with value "0" in the dataset to fill the missing values. Another issue is the presence of corrupted data, marked by a "−1" in the dataset, which are relative to periods when the telephone activity is not registered correctly. To address this problem, we followed the procedure described by Shen and Huang (2005) and we replaced the corrupted entries with the average value of the corresponding periods (same weekday and hour) from the two adjacent weeks. Contrarily to some other works on STLF (Ibrahim and L'Ecuyer 2013; Shen and Huang 2008; Andrews and Cunningham 1995), we decided to not discard outliers, such as holidays or days with an anomalous number of calls, or we modeled them as separate variables.

As next step in our pre-analysis, we identify the main seasonality in the data. We analyze $ts1$, but similar considerations hold also for the remaining time series. Through frequency analysis and by inspecting the autocorrelation function, depicted as a gray line in Fig. 6.1b, it emerges a strong seasonal pattern every 24 hours. As expected, data experience regular and predictable daily changes, due to the nature of the telephonic traffic. This cycle represents the main seasonality and we filter it out by applying a seasonal differencing with lag 24. In this way, the RNNs focus on learning to predict the series of changes in each seasonal cycle. The practice of removing the seasonal effect from the time series demonstrated to improve the prediction accuracy of models based on neural networks (Zhang and Kline 2007; Claveria et al. 2017). The black line in Fig. 6.1b depicts the autocorrelation of the time series after seasonal differentiation. Except from the high anticorrelation at lag 24, introduced by the differentiation, the time series appears to be uncorrelated elsewhere and, therefore, we can exclude the presence of the second, less obvious seasonality.

Due to the nature of the seasonality in the data, we expect a strong relationship between the time series of the loads ($ts1$–$ts4$) and $ts5$, which is relative to the hour of the day. On the other hand, we envisage a lower dependency of the loads with $ts6$, the time series of the weekdays, since we did not notice the presence of the second seasonal cycle after the differentiation at lag 24. To confirm our hypothesis, we computed the mutual information between the time series, which are reported in the Hinton diagram in Fig. 6.2. The size of the blocks is proportional to the degree of mutual information among the time series. Due to the absence of strong relationships, we decided to discard $ts6$ to reduce the complexity of the model by excluding a variable with potentially low impact in the prediction task. We also discarded $ts5$ because the presence of the cyclic daily pattern is already accounted by doing the seasonal differencing at lag 24. Therefore, there is no need to provide daily hours as an additional exogenous input.

Besides differentiation, a common practice in STLF is to apply some form of normalization to the data. We applied a standardization (z-score) but rescaling into the interval $[−1, 1]$ or $[0, 1]$ are other viable options. Additionally, a nonlinear transformation of the data by means of a nonlinear function (e.g., square root or logarithm) can remove some kinds of trend and stabilize the variance in the data, without altering too much their underlying structure (Weinberg et al. 2007; Shen and Huang 2008; Ibrahim and L'Ecuyer 2013). In particular, a log-transform is suitable for a set of

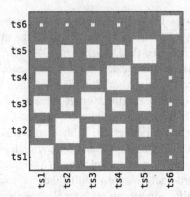

Fig. 6.2 Hinton diagram of the mutual information between the time series in the Orange dataset. The size of each block is proportional to the degree of mutual information among the time series. The measurement indicates a strong relationship between the load time series and the daily hours (ts5), while the dependency with the day of the week (ts6) is low

random variables characterized by a high variability in their statistical dispersion (heteroscedasticity) or for a process whose fluctuation of the variance is larger than the fluctuation of the mean (overdispersion). To check those properties, we analyze the mean and the variance of the telephonic traffic within the main seasonal cycle across the whole dataset. The solid black line in Fig. 6.3a represents the mean load of ts1, while the shaded gray area illustrates the variance. As we can see, the data are not characterized by overdispersion, since the fluctuations of the mean are greater than the ones of the variance. However, we notice the presence of heteroscedasticity, since the amount of variance changes in different hours of the day. In fact, the central hours where the amount of telephonic activity is higher are characterized by a greater standard deviation in the load. In Fig. 6.3b, we observe that by applying a log-transform we significantly reduce the amount of variance in the periods characterized by a larger traffic load. However, after the log-transformation, the mean value of the load becomes more flattened and the variance relative to periods with lower telephonic activity is enhanced. This could cause issues during the training of the RNN, and hence in the experiments, we evaluate the prediction accuracy both with and without applying the log-transformation to the data.

Preprocessing transformations are applied in this order: (i) log-transform, (ii) seasonal differencing at lag 24, and (iii) standardization. Each preprocessing operation is successively reversed to evaluate the forecast produced by each RNN.

6.2 ACEA Dataset—Electricity Load

The second time series we analyze is relative to the electricity consumption registered by ACEA (Azienda Comunale Energia e Ambiente), the company which provides the electricity to Rome and some neighboring regions. The ACEA power grid in

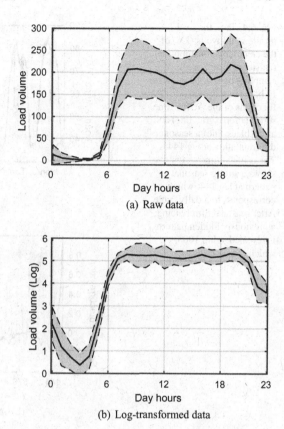

Fig. 6.3 Average weekly load (solid black line) and the standard deviation (shaded gray area) of the telephonic activity in the whole dataset

(a) Raw data

(b) Log-transformed data

Rome consists of 10.490 Km of medium voltage lines, while the low-voltage section covers 11.120 km. The distribution network is constituted of backbones of uniform section, exerting radially and with the possibility of counter supply if a branch is out of order. Each backbone is fed by two distinct primary stations and each half-line is protected against faults through the breakers. Additional details can be found in Ref. Santis et al. (2015). The time series we consider concerns the amount of supplied electricity, measured on a medium voltage feeder from the distribution network of Rome. Data are collected every 10 min for 954 days of activity (almost 3 years), spanning from 2009 to 2011, for a total of 137444 measurements. Also, in this case, we train the RNNs to predict the electricity load 24 h ahead, which corresponds to 144 time steps ahead prediction. For this forecast task, we do not provide any exogenous time series to the RNNs. In the hyperparameter optimization, we use the load relative to the first 3 months as training set and the load of the fourth month as validation set. Once the best hyperparameter configuration is identified, we fine-tune each RNN on the first 4 months and we use the fifth month as test set to evaluate and to compare the accuracy of each network.

Fig. 6.4 In **a**, the load profile in kilovolts (kV) of the electricity consumption registered over 1 week. The sampling time is 10 minutes. In **b**, the autocorrelation functions of the ACEA time series before (gray line) and after (black line) a seasonal differentiation at lag 144. The original time series shows a strong seasonal pattern at lag 144, which corresponds to a daily cycle. After seasonal differencing, a previously hidden pattern is revealed at lag 1008, which corresponds to a weekly cycle

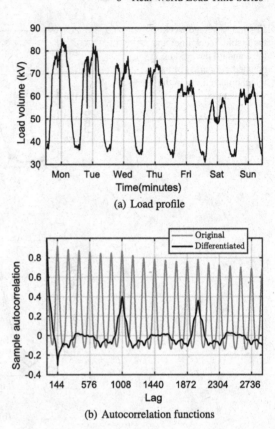

(a) Load profile

(b) Autocorrelation functions

A profile of the electric consumption over 1 week (1008 measurements) is depicted in Fig. 6.4a.

In the ACEA time series, there are no missing values but 742 measurements (which represent 0.54% of the whole dataset) are corrupted. The consumption profile is more irregular in this time series, with respect to the telephonic data from the Orange dataset. Therefore, rather than replacing the corrupted values with an average load, we used a form of imputation with a less strong bias. Specifically, we first fit a cubic spline to the whole dataset and then we replaced the corrupted entries with the corresponding values from the fitted spline. In this way, the imputation better accounts for the local variations of the load.

Also in this case, we perform a preemptive analysis in order to understand the nature of the seasonality, to detect the presence of hidden cyclic patterns, and to evaluate the amount of variance in the time series. By computing the autocorrelation function up to a sufficient number of lags, depicted as a gray line in Fig. 6.4b, it emerges a strong seasonality pattern every 144 time intervals. As expected, this corresponds exactly to the number of measurements in 1 day. By differencing the time series at lag 144, we remove the main seasonal pattern and the trend. Also,

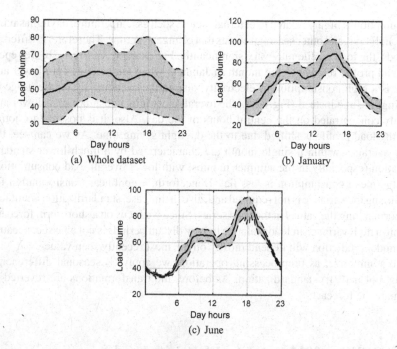

Fig. 6.5 In **a**, we report the mean load (black line) and the standard deviation (gray area) of the electricity consumption in a week, accounting the measurements from all the dataset. In **b** and **c**, the measurements are relative only to 1 month of activity, which are January and June, respectively

in this case, the negative peak at lag 144 is introduced by the differentiation. If we observe the autocorrelation plot of the time series after seasonal differencing (black line in Fig. 6.4b) we observe a second strong correlation at lag 1008. This second seasonal pattern represents a weekly cycle that was not clearly visible before the differentiation. Due to the long periodicity of the time cycle, to account this second seasonality, a predictive model would require a large amount of memory to store information for a longer time interval. While a second differentiation can remove this second seasonal pattern, we would have to discard the values relative to the last week of measurements. Most importantly, the models we train could not learn the similarities in consecutive days at a particular time, since they would be trained on the residuals of the load at the same time and day in two consecutive weeks. Therefore, we decided to apply only the seasonal differentiation at lag 144.

To study the variance in the time series, we consider the average daily load over the main seasonal cycle of 144 time intervals. As we can see from Fig. 6.5a, data appear to be affected by overdispersion, as the standard deviation (gray-shaded areas) fluctuates more than the mean. Furthermore, the mean load value (black solid line) seems to not change much across the different hours, while it is reasonable to expect significant differences in the load between night and day. However, we remind that the Acea time series spans a long time lapse (almost 3 years) and that the electric consumption is highly related to external factors such as temperature, daylight saving time, holidays, and other seasonal events that change over time. Therefore, in different periods, the load profile may vary significantly. For example, in Fig. 6.5b, we report the load profile relative to the month of January, when temperatures are lower and there is a high consumption of electricity, also in the evening, due to the usage of heating. In June instead (Fig. 6.5c), the overall electricity consumption is lower and mainly concentrated on the central hours of the day. Also, it is possible to notice that the load profile is shifted due to the daylight saving time. As we can see, the daily averages within a single month are characterized by a much lower standard deviation (especially in the summer months, with lower overall load consumption) and the mean consumption is less flat. Henceforth, a nonlinear transformation for stabilizing the variance is not required and, also in this case, standardization is suitable for normalizing the values in the time series. Since we focus on a short-term forecast, having a high variance in loads relative to very distant periods is not an issue, because the model prediction will depend mostly on the most recently seen values.

To summarize, as preprocessing operation, we apply (i) seasonal differencing at lag 144 and (ii) standardization. As before, the transformations are reverted to estimate the forecast.

6.3 GEFCom2012 Dataset—Electricity Load

The last real-world dataset that we study is the time series of electricity consumption from the Global Energy Forecasting Competition (GEFCom) 2012) (Kaggle 2012). The GEFCom 2012 dataset consists of 4 years (2004–2007) of hourly electricity load collected from a US energy supplier. The dataset comprises time series of consumption measurements, from 20 different feeders in the same geographical area. The values in each time series represent the average hourly load, which varies from 10.000kWh to 200.000kWh. The dataset also includes time series of the temperatures registered in the area where the electricity consumption is measured.

The forecast task that we tackle is the 24 hours ahead prediction of the aggregated electricity consumption, which is the sum of the 20 different load time series in the year 2006. The measurements relative to the to first 10 months of the 2006 are used as training set, while the eleventh month is used as validation set for guiding the hyperparameters optimization. The time series of the temperature in the area is also provided to the RNNs as an exogenous input. The prediction accuracy of the optimized RNNs is then evaluated on the last month of the 2006. A depiction of the load profile of the aggregated load time series is reported in Fig. 6.6a. We can observe a trend in the time series, which indicates a decrement in the energy demand over time. This can be related to climate conditions since, as the temperature becomes warmer during the year, the electricity consumption for the heating decreases.

To study the seasonality in the aggregated time series, we evaluate the autocorrelation function, which is depicted as the gray line in Fig. 6.6b. From the small subplot in top-right part of the figure, relative to a small segment of the time series, it emerges

Fig. 6.6 In **a**, the load profile in kilowatt hour (kWh) of the aggregated electricity consumption registered in the first 4 months of activity in 2006, from the GEFCom dataset. The sampling time in the time series is 1 hour. In **b**, the autocorrelation functions of the GEFCom time series before (gray line) and after (black line) a seasonal differentiation at lag 24. The small subplot on the top-right part of the figure reports a magnified version of the autocorrelation function before differentiation at lag $t = 200$

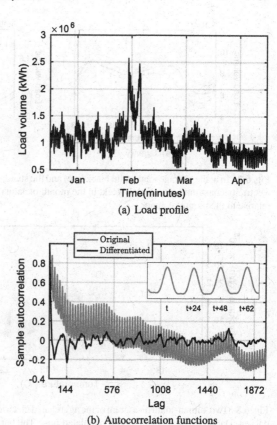

(a) Load profile

(b) Autocorrelation functions

a strong seasonal pattern every 24 hours. By applying a seasonal differentiation with lag 24, the main seasonal pattern is removed, as we can see from the autocorrelation function of the differentiated time series, depicted as a black line in the figure. After differentiation, the autocorrelation becomes close to zero after the first lag and, therefore, we can exclude the presence of the second, strong seasonal pattern (e.g., a weekly pattern).

Similarly to what we did previously, we analyze the average load of the electricity consumption during 1 week. As for the ACEA dataset, rather than considering the whole dataset, we analyze separately the load 1 month in winter and 1 month in summer. In Fig. 6.7a, we report the mean load (black line) and standard deviation (gray area) in January. Figure 6.7b instead depicts the measurements for May. It is possible to notice a decrement of the load during the spring period, due to the reduced usage of heating. It is also possible to observe a shift in the consumption profile to later hours in the day, due to the time change. By analyzing the amount of variance and the fluctuations of the mean load, we can exclude the presence of overdispersion and heteroscedasticity phenomena in the data.

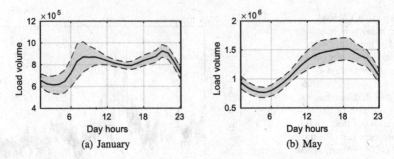

Fig. 6.7 In **a**, the average load (solid black line) and the standard deviation (shaded gray area) of the electricity consumption during 1 week, in the month of January. In **b**, we report the measurements relative to the month of June

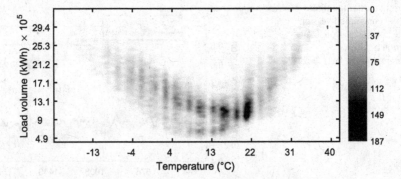

Fig. 6.8 Two-dimensional histogram of the aggregated electricity load and temperature in GEFCom dataset. Darker areas represent more populated bins. The bar on the right indicates the number of elements in each bin. The characteristic V-shape of the resulting pattern is because of the increased use of heating and cooling devices in the presence of hot and cold temperatures

To improve the forecasting accuracy of the electricity consumption, a common practice is to provide to the prediction system the time series of the temperature as an exogenous variable. In general, the load and the temperature are highly related, since both in the coldest and warmest months electricity demand increases, due to the usage of heating and air conditioning, respectively. However, the relationship between temperature and load cannot be captured by the linear correlation, since the consumption increases both when temperatures are too low or too high. Indeed, the estimated correlation between the aggregated load time series of interest and the time series of the temperature in the area yields only a value of 0.2. However, their relationship is evidenced by computing a two-dimensional histogram of the two variables, proportional to their estimated joint distribution, which is reported in Fig. 6.8. The V-shape denotes an increment of the electricity consumption for low and high temperatures with respect to a mean value of about 22° C.

The preprocessing operations we apply on the GEFCom dataset are (i) seasonal differencing at lag 24 and (ii) standardization. Also, in this case, these transformations are reverted to estimate the forecast.

References

Andrews BH, Cunningham SM (1995) LL Bean improves call-center forecasting. Interfaces 25(6):1–13

Bianchi FM, Rizzi A, Sadeghian A, Moiso C (2016) Identifying user habits through data mining on call data records. Eng Appl Artif Intell 54:49–61. https://doi.org/10.1016/j.engappai.2016.05.007

Blondel VD, Esch M, Chan C, Clérot F, Deville P, Huens E, Morlot F, Smoreda Z, Ziemlicki C (2012) Data for development: the D4D challenge on mobile phone data. arXiv:12100137

Claveria O, Monte E, Torra S (2017) Data pre-processing for neural network-based forecasting: does it really matter? Technological and Economic Development of Economy 0(0):1–17. https://doi.org/10.3846/20294913.2015.1070772

Franses PH (1991) Seasonality, non-stationarity and the forecasting of monthly time series. Int J Forecast 7(2):199–208

Ibrahim R, L'Ecuyer P (2013) Forecasting call center arrivals: Fixed-effects, mixed-effects, and bivariate models. Manuf Serv Oper Manag 15(1):72–85

Kaggle (2012) GEFCom 2012 global energy forecasting competition 2012. https://www.kaggle.com/c/global-energy-forecasting-competition-2012-load-forecasting. Accessed 26 April 2017

Orange (2013) D4d challenge. https://www.d4d.orange.com/en/Accueil. Accessed 22 Sep 2016

Santis ED, Livi L, Sadeghian A, Rizzi A (2015) Modeling and recognition of smart grid faults by a combined approach of dissimilarity learning and one-class classification. Neurocomputing 170:368–383. https://doi.org/10.1016/j.neucom.2015.05.112. (Advances on Biological Rhythmic Pattern Generation: Experiments, Algorithms and Applications Selected Papers from the 2013 International Conference on Intelligence Science and Big Data Engineering (IScIDE 2013)Computational Energy Management in Smart Grids)

Shen H, Huang JZ (2005) Analysis of call centre arrival data using singular value decomposition. Appl Stoch Models Bus Ind 21(3):251–263

Shen H, Huang JZ (2008) Interday forecasting and intraday updating of call center arrivals. Manuf Serv Oper Manag 10(3):391–410

Weinberg J, Brown LD, Stroud JR (2007) Bayesian forecasting of an inhomogeneous Poisson process with applications to call center data. J Am Stat Assoc 102(480):1185–1198

Zhang GP, Kline DM (2007) Quarterly time-series forecasting with neural networks. IEEE Trans Neural Netw 18(6):1800–1814. https://doi.org/10.1109/TNN.2007.896859

Zhang G, Qi M (2005) Neural network forecasting for seasonal and trend time series. Eur J Oper Res 160(2):501–514. https://doi.org/10.1016/j.ejor.2003.08.037. (Decision support systems in the internet age)

Chapter 7
Experiments

Abstract In this section, we compare the prediction performance achieved by the recurrent neural network architectures presented in the previous sections on both the synthetic tasks and the real-world datasets. For each architecture, we report the optimal configuration of its hyperparameters for the task at hand, and the best learning strategy adopted for training the model weights. We perform several independent evaluation of the prediction results due to the stochastic initialization of the internal model weights. The accuracy of the forecast is evaluated in terms of normalized mean squared error and the results are reported both as numerical value and graphical depictions of the predicted time series.

Keywords Prediction accuracy evaluation · Performance analysis · Forecast models comparative analysis

During the validation phase, different configurations are randomly selected from admissible intervals and, once the training is over, their optimality is evaluated as the prediction accuracy achieved on the validation set. We opted for a random search as it can find more accurate results than a grid search, when the same number of configurations is evaluated (Bergstra and Bengio 2012). Once the (sub)optimal configuration is identified, we train each model 10 times on the training and validation data, using random and independent initializations of the network parameters, and we report the highest prediction accuracy obtained on the unseen values of the test set.

To compare the forecast capability of each model, we evaluate the prediction accuracy ψ as $\psi = 1 - $ NRMSE. NRMSE is the normalized root mean squared error that reads

$$\text{NRMSE}\left(\mathscr{Y}, \mathscr{Y}^*\right) = \sqrt{\frac{\langle \|\mathscr{Y} - \mathscr{Y}^*\|^2 \rangle}{\langle \|\mathscr{Y} - \langle \mathscr{Y}^* \rangle\|^2 \rangle}}, \qquad (7.1)$$

where $\langle \cdot \rangle$ computes the mean, $\mathscr{Y}$ are the RNN outputs and $\mathscr{Y}^*$ are the ground truth values.

In the following, we present two types of experiments. The first experiment consists in the prediction of the synthetic time series presented in Chap. 5, commonly considered as benchmarks in forecast applications, and the results are discussed in Sect. 7.2. In the second experiment, we forecast the real-world telephonic and

F. M. Bianchi et al., *Recurrent Neural Networks for Short-Term Load Forecasting*,
SpringerBriefs in Computer Science, https://doi.org/10.1007/978-3-319-70338-1_7

electricity load time series, presented in Chap. 6 The results of this second experiment are discussed in Sect. 7.3.

7.1 Experimental Settings

7.1.1 ERNN, LSTM, and GRU

The three RNNs described in Chap. 3 have been implemented in Python, using Keras library with Theano Theano Development Team (2016) as backend.[1]

To identify an optimal configuration for the specific task at hand, we evaluate for each RNN different values of the hyperparameters and training procedures. The configurations are selected randomly and their performances are evaluated on the validation set, after having trained the network for 400 epochs. To get rid of the initial transient phase, we drop the first 50 outputs of the network. A total of 500 random configurations for each RNN are evaluated and, once the optimal configuration is found, we compute the prediction accuracy on the test set. In the test phase, each network is trained for 2000 epochs.

The optimization is performed by assigning to each hyperparameter a value uniformly sampled from a given interval, which can be continuous or discrete. The gradient descent strategies are selected from a set of possible alternatives, which are SGD, Nesterov momentum, and Adam. For SGD and Nesterov, we anneal the learning rate with a step decay of 10^{-6} in each epoch. The learning rate η is sampled from different intervals, depending on the strategy selected. Specifically, for SGD, we set $\eta = 10^c$, with c uniformly sampled in $[-3, -1]$. For Nesterov and Adam, since they benefit from a smaller initial value of the learning rate, we sample c uniformly in $[-4, -2]$. The remaining hyperparameters used in the optimization strategies are kept fixed to their default values (see Sect. 2.3). Regarding the number N_h of hidden units in the recurrent hidden layer, we randomly chose for each architecture four possible configurations that yield an amount of trainable parameters approximately equal to 1800, 3900, 6800, and 10000. This corresponds to $N_h = \{40, 60, 80, 100\}$ in ERNN, $N_h = \{20, 30, 40, 50\}$ in LSTM and $N_h = \{23, 35, 46, 58\}$ in GRU. For each RNNs, N_h is randomly selected from these sets. To deal with the problem of vanishing gradient discussed in Sect. 2.4, we initialize the RNN weights by sampling them from an uniform distribution in $[0, 1]$ and then rescaling their values by $1/\sqrt{N_h}$. For the L_1 and L_2 regularization terms, we sample independently λ_1 and λ_2 from $[0, 0.1]$, an interval containing values commonly assigned to these hyperparameters in RNNs Zeyer et al. (2016). We apply the same regularization to input, recurrent, and output weights. As suggested by Gal and Ghahramani (2015), we drop the same input and recurrent connections at each time step in the BPTT, with a dropout prob-

---

[1] Keras library is available at https://github.com/fchollet/keras. Theano library is available at http://deeplearning.net/software/theano/.

ability p_{drop} drawn from $\{0, 0.1, 0.2, 0.3, 0.5\}$, which are commonly used values (Pham et al. 2014b). If $p_{drop} \neq 0$, we also apply a L_2 regularization. This combination usually yields a lowest generalization error than dropout alone (Srivastava et al. 2014). Note that another possible approach combines dropout with the max-norm constraint, where the L_2 norm of the weights is clipped whenever it grows beyond a given constant, which however, introduces another hyperparameter.

For the training, we consider the backpropagation through time procedure BPTT(τ_b, τ_f) with $\tau_b = 2\tau_f$. The parameter τ_f is randomly selected from the set $\{10, 15, 20, 25, 30\}$. As we discussed in Sect. 2.1, this procedure differs from both the *true* BPTT and the *epochwise* BPTT (Williams and Zipser 1995), which is implemented as default by popular deep learning libraries such as TensorFlow (Abadi et al. 2015).

7.1.2 NARX

This RNN is implemented using the MATLAB neural network toolbox.[2] We configured NARX network with an equal number of input and output lags on the TDLs ($d_x = d_y$) and with the same number of neurons N_h in each one of the N_l hidden layers. Parameters relative to weight matrices and bias values $\Theta = \{\theta, \theta_o, \theta_{h_1}, \ldots, \theta_{h_{N_l}}\}$ are trained with a variant of the quasi-Newton search, called Levenberg–Marquardt optimization algorithm. This is an algorithm for error backpropagation that provides a good tradeoff between the speed of the Newton algorithm and the stability of the steepest descent method (Battiti 1992). The loss function to be minimized is defined in Eq. 4.6.

NARX requires the specification of 5 hyperparameters, which are uniformly drawn from different intervals. Specifically, TDL lags are drawn from $\{2, 3, \ldots, 10\}$; the number of hidden layers N_l is drawn from $\{1, 2, \ldots, 5\}$; the number of neurons N_h in each layer is drawn from $\{5, 6, \ldots, 20\}$; the regularization hyperparameter λ_2 in the loss function is randomly selected from $\{2^{-1}, 2^{-2}, \ldots, 2^{-10}\}$; the initial value η of learning rate is randomly selected from $\{2^{-5}, 2^{-6}, \ldots, 2^{-25}\}$.

A total of 500 random configurations for NARX are evaluated and, for each hyperparameters setting, the network is trained for 1000 epochs in the validation. In the test phase, the network configured with the optimal hyperparameters is trained for 2000 epochs. Also in this case, we discard the first 50 network outputs to get rid of the initial transient phase of the network.

---

[2]https://se.mathworks.com/help/nnet/ref/narxnet.html.

7.1.3   ESN

For the ESN, we used a modified version of the Python implementation,[3] provided by Løkse et al. (2017). Learning in ESN is fast, as the readout is trained by means of a linear regression. However, the training does not influence the internal dynamics of the random reservoir, which can be controlled only through the ESN hyperparameters. This means that a more accurate (and computationally intensive) search of the optimal hyperparametyers is required with respect to the other RNN architectures. In RNNs, the precise, yet slow gradient-based training procedure is mainly responsible for learning the necessary dynamics and it can compensate a suboptimal choice of the hyperparameters.

Therefore, in the ESN validation phase, we evaluate a larger number of configurations (5000), by uniformly drawing 8 different hyperparameters from specific intervals. In particular, the number of neurons in the reservoir, N_h, is drawn from $\{400, 450, \ldots, 900\}$; the reservoir spectral radius, ρ, is drawn in the interval $[0.5, 1.8]$; the reservoir connectivity R_c is drawn from $[0.15, 0.45]$; the noise term ξ in Eq. (4.7) comes from a Gaussian distribution with zero mean and variance drawn from $[0, 0.1]$; scaling of input signal ω_i and desired response ω_o are drawn from $[0.1, 1]$; scaling of output feedback ω_f is drawn from $[0, 0.5]$; the linear regression regularization parameter λ_2 is drawn from $[0.001, 0.4]$. Also in this case, we discarded the first 50 ESN outputs relative to the initial transient phase.

7.2   Results on Synthetic Dataset

In Fig. 7.1, we report the prediction accuracy obtained by the RNNs on the test set of the three synthetic problems. The best configurations of the architectures identified for each task through random search are reported in Table 7.1.

First of all, we observe that the best performing RNN is different in each task. In the MG task, ESN outperforms the other networks. This result confirms the excellent and well-known capability of the ESN in predicting chaotic time series (Li et al 2012b; Jaeger and Haas 2004). In particular, ESN demonstrated to be the most accurate architecture for the prediction of the MG system (Shi and Han 2007). The ESN achieves the best results also in the MSO task, immediately followed by ERNN. On the NARMA task, instead, ESN performs poorly, while the LSTM is the RNN that predicts the target signal with the highest accuracy.

In each test, NARX struggles in reaching performance comparable with the other architectures. In particular, in NARMA and MSO task, the NRMSE prediction error of NARX is 0.53 and 1.99, respectively (note that we cut the y-axis to better show the remaining bars). Note that, since the NRMSE is normalized by the variance of the target signal, an error greater than 1 means that the performance is worse than a constant predictor, with value equal to the mean of the target signal.

---

[3]https://github.com/siloekse/PythonESN.

Fig. 7.1 NRMSE values
achieved on the test sets by
each RNN architecture on
the three synthetic prediction
tasks

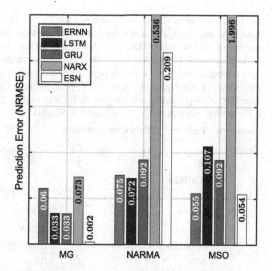

It is also interesting to notice that in MSO, ERNN achieves a prediction accuracy
higher than GRU and LSTM. Despite the fact that the MSO task demands a large
amount of memory, due to the extremely long periodicity of the target signal, the two
gated architectures (LSTM and GRU) are not able to outperform the ERNN. We can
also notice that for MSO, the optimal number of hidden nodes (N_h) is lower than in
the other tasks. A network with a limited complexity is less prone to overfit on the
training data, but it is also characterized by an inferior modeling capability. Such a
high modeling capability is not needed to solve the MSO task, given that the network
manages to learn correctly the frequencies of the superimposed sinusoidal signals.

Finally, we observe that LSTM and GRU perform similarly on the each task, but
there is not a clear winner. This finding is in agreement with previous studies, which,
after several empirical evaluations, concluded that it is difficult to choose in advance
the most suitable gated RNN to solve a specific problem (Chung et al. 2014).

Regarding the gradient descent strategies used to train the parameters in RNN,
LSTM, and GRU, we observe in Table. 7.1 that Adam is often identified as the
optimal strategy. The standard SGD is selected only for GRU in the MG task. This
is probably a consequence of the lower convergence rate of the SGD minimization,
which struggles to discover a configuration that achieves a good prediction accuracy
on the validation set in the limited amount (400) of training epochs. Also, the Nesterov
approach seldom results to be as the optimal strategy and a possible explanation is
its high sensitivity to the (randomly selected) learning rate. In fact, if the latter is
too high, the gradient may build up too much momentum and bring the weights into
a configuration where the loss function is very large. This results in even greater
gradient updates, which leads to rough oscillations of the weights that can reach very
large values.

From the optimal configurations in Table 7.1, another striking behavior about the
optimal regularization procedures emerges. In fact, we observe that in each RNN

Table 7.1 Optimal RNNs configurations for solving the three synthetic prediction tasks, MG, NARMA, and MSO. The acronyms in the table are N_h—number of nodes in the hidden layer; N_l – number of hidden layers; TDL—number of lags on the tapped delay lines; η—learning rate; λ_1— L_1 regularization parameter; λ_2—L_2 regularization parameter; OPT—gradient descent strategy; τ_f—number of new time steps processed before computing the BPTT; τ_b—number of time step the gradient is propagated back in BPTT; p_{drop}—dropout probability; ρ—spectral radius of ESN reservoir; R_c—percentage of sparsity in ESN reservoir; ξ—noise in ESN state update; ω_i, ω_o, ω_f—scaling of input, teacher and feedback weights

Network	Task	RNN Configuration							
Narx		N_h		N_l		TDL		η	λ_2
	MG	15		2		6		3.8E-6	0.0209
	NARMA	17		2		10		2.4E-4	0.4367
	MSO	12		5		2		0.002	0.446
ERNN		τ_b	τ_f	N_h	OPT	η	p_{drop}	λ_1	λ_2
	MG	20	10	80	Adam	0.00026	0	0	0.00037
	NARMA	50	25	80	Nesterov	0.00056	0	0	1E-5
	MSO	50	25	60	Adam	0.00041	0	0	0.00258
LSTM		τ_b	τ_f	N_h	OPT	η	p_{drop}	λ_1	λ_2
	MG	50	25	40	Adam	0.00051	0	0	0.00065
	NARMA	40	20	40	Adam	0.00719	0	0	0.00087
	MSO	50	25	20	Adam	0.00091	0	0	0.0012
GRU		τ_b	τ_f	N_h	OPT	η	p_{drop}	λ_1	λ_2
	MG	40	20	46	SGD	0.02253	0	0	6.88E-6
	NARMA	40	20	46	Adam	0.00025	0	0	0.00378
	MSO	50	25	35	Adam	0.00333	0	0	0.00126
ESN		N_h	ρ	R_c	ξ	ω_i	ω_o	ω_f	λ_2
	MG	800	1.334	0.234	0.001	0.597	0.969	0.260	0.066
	NARMA	700	0.932	0.322	0.013	0.464	0.115	0.045	0.343
	MSO	600	1.061	0.231	0.002	0.112	0.720	0.002	0.177

and for each task, only the L_2 norm of the weights is the optimal regularizer. On the other hand, the parameters λ_1 and p_{drop} relative to the L_1 norm and the dropout are always zero. This indicates that to successfully solve the synthetic prediction tasks, it is sufficient to train the networks with small weights in order to prevent the overfitting.

Finally, we notice that the best results are often found using network with a high level of complexity, in terms of number of neurons and long windows in BPTT or TDL, for Narx. In fact, in most cases, the validation procedure identifies the optimal values for these variables to be close to the upper limit of their admissible intervals. This is somehow expected, since a more complex model can achieve higher modeling capabilities, if equipped with a suitable regularization procedure to prevent overfitting during training. However, the tradeoff in terms of computational resources for training more complex models is often very high and small increments in the performance are obtained at the cost of much longer training times.

7.3   Results on Real-World Dataset

The highest prediction accuracies obtained by the RNNs on the test set (unseen data) of the real-world load time series are reported in Fig. 7.2. As before, in Table 7.2, we report the optimal configuration of each RNN for the different tasks.

7.3.1   Results on Orange Dataset

All the RNNs achieve very similar prediction accuracy on this dataset, as it is possible to see from the first bar plot in Fig. 7.2. In Fig. 7.3, we report the residuals, depicted as black areas, between the target time series and the forecasts of each RNN. The figure gives immediately a visual quantification of the accuracy, as the larger the black areas, the greater the prediction error in that parts of the time series. In particular, we observe that the values which the RNNs fail to predict are often relative to the same interval. Those values represent fluctuations that are particularly hard to forecast, since they correspond to unusual increments (or decrements) of load, which differ significantly from the trend observed in the past. For example, the error increases when the load suddenly grows in the last seasonal cycle in Fig. 7.3.

Fig. 7.2 NRMSE values achieved on the test sets by each RNN architecture on the three real-world STLF problems. Note that scales are different for each dataset

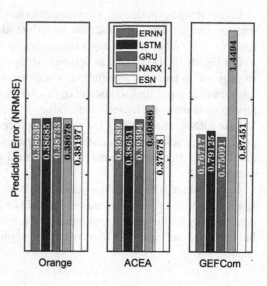

Table 7.2 Optimal RNNs configurations adopted in the three real-world STLF problems. Refer to Table 7.1 for the definition of the acronyms in this table.

Network	Task	RNN Configuration							
Narx		N_h		N_l		TDL		η	λ_2
	Orange	11		4		2		$1.9E-6$	0.082
	ACEA	11		3		2		$1.9E-6$	0.0327
	GEFCom	18		4		9		$6.1E-5$	0.3136
ERNN		τ_b	τ_f	N_h	OPT	η	p_{drop}	λ_1	λ_2
	Orange	30	15	100	SGD	0.011	0	0	0.0081
	ACEA	60	30	80	Nesterov	0.00036	0	0	0.0015
	GEFCom	50	25	60	Adam	0.0002	0	0	0.0023
LSTM		τ_b	τ_f	N_h	OPT	η	p_{drop}	λ_1	λ_2
	Orange	40	20	50	Adam	0.0013	0	0	0.0036
	ACEA	50	25	40	Adam	0.0010	0.1	0	0.0012
	GEFCom	50	25	20	SGD	0.0881	0	0	0.0017
GRU		τ_b	τ_f	N_h	OPT	η	p_{drop}	λ_1	λ_2
	Orange	40	20	46	SGD	0.0783	0	0.0133	0.0004
	ACEA	40	20	35	Adam	0.0033	0	0	0.0013
	GEFCom	60	30	23	Adam	0.0005	0	0	0.0043
ESN		N_h	ρ	R_c	ξ	ω_i	ω_o	ω_f	λ_2
	Orange	400	0.5006	0.3596	0.0261	0.2022	0.4787	0.1328	0.3240
	ACEA	800	0.7901	0.4099	0.0025	0.1447	0.5306	0.0604	0.1297
	GEFCom	500	1.7787	0.4283	0.0489	0.7974	0.9932	0.0033	0.2721

In the Orange experiment, we evaluate the results with or without applying a log-transform to the data. We observed sometime log-transform yields slightly worse result ($\sim 0.1\%$), but in most cases the results are equal.

For ERNN, SGD is found as optimal, which is a slower yet more precise update strategy and is more suitable for gradient descent if the problem is difficult. ERNN takes into account a limited amount of past information, as the window in the BPTT procedure is set to a relatively small value.

Like ERNN, also for GRU, the validation procedure identified SGD as the optimal gradient descent strategy. Interestingly, L_1 regularization is used, while in all the other cases it is not considered. On the other hand, the L_2 regularization parameter is much smaller.

In the optimal NARX configuration, TDL is set to a very small value. In particular, since the regression is performed only on the last 2 time intervals, the current output depends only on the most recent inputs and estimated outputs. From the number of hidden nodes and layers, we observe that the optimal size of the network is relatively small.

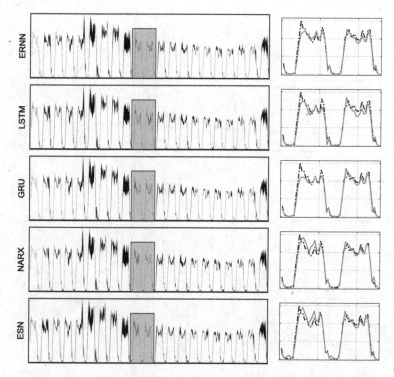

Fig. 7.3 Orange dataset—The plots on the left show the residuals of predictions of each RNN with respect to the ground truth; black areas indicate the errors in the predictions. The plots on right depict a magnification of the area in the gray boxes from the left graphics; the dashed black line is the ground truth, the solid gray line is the prediction of each RNN

Relatively to the ESN configuration, we notice a very small spectral radius. This means that, also in this case, the network is configured with a small amount of memory. This results in reservoir dynamics that are more responsive and, consequently, in outputs that mostly depend on the recent inputs. As a consequence, the value of input scaling is small, since there is no necessity of quickly saturating the neurons activation.

7.3.2 Results on ACEA Dataset

The time series of the electricity load is quite regular except for few, erratic fluctuations. As for the Orange dataset, RNN predictions are inaccurate mainly in correspondence of such fluctuations, while they output a correct prediction elsewhere. This behavior is outlined by the plots in Fig. 7.4, where we observe that the residuals are small and, in each RNN prediction, they are mostly localized in common

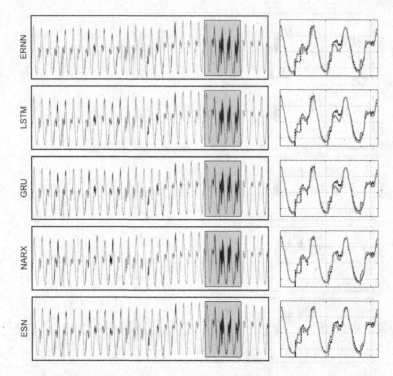

Fig. 7.4 ACEA dataset—The plots on the left show the residuals of predictions of each RNN with respect to the ground truth; black areas indicate the errors in the predictions. The plots on right depict a magnification of the area in the gray boxes from the left graphics; the dashed black line is the ground truth, the solid gray line is the prediction of each RNN

time intervals. From the NRMSE values in Fig. 7.2, we see that ESN performs better than the other networks. The worst performance is obtained by NARX, while the gradient-based RNNs yield better results, which are very similar to each other.

In ERNN and GRU, the optimal regularization found is the L_2 norm, whose coefficient assumes a small value. In LSTM instead, besides the L_2 regularization term, the optimal configuration includes also a dropout regularization with a small probability. The BPTT windows have comparable size in all the gradient-based networks.

The optimal NARX configuration for ACEA is very similar to the one identified for Orange and is characterized by a low complexity in terms of number of hidden nodes and layers. Also in this case, the TDLs are very short.

Similarly to the optimal configuration for Orange, the ESN spectral radius assumes a small value, meaning that the network is equipped with a short-term memory and it captures only short temporal correlations in the data. The reservoir is configured with a high connectivity, which yields more homogeneous internal dynamics.

7.3.3 Results on GEFCom Dataset

This time series is more irregular than the previous ones, as it shows a more noisy behavior that is harder to be predicted. From Fig. 7.5, we see that the extent of the black areas of the residual is much larger than in the other datasets, denoting a higher prediction error. From the third panel in Fig. 7.2, we observe larger differences in the results with respect to the previous cases. In this dataset, the exogenous time series of temperature plays a key role in the prediction, as it conveys information that are particularly helpful to yield a high accuracy. The main reason of the discrepancy in the results for the different networks may be in their capability of correctly leveraging this exogenous information for building an accurate forecast model.

From the results, we observe that the gradient-based RNNs yield the best prediction accuracy. In particular, ERNN and GRU generate a prediction with the lowest NRMSE with respect to the target signal. ESN, instead, obtains considerably lower performance. Like for the syntethic datasets NARMA and MSO, NARX produces a very inaccurate prediction, scoring a NRMSE which is above 1.

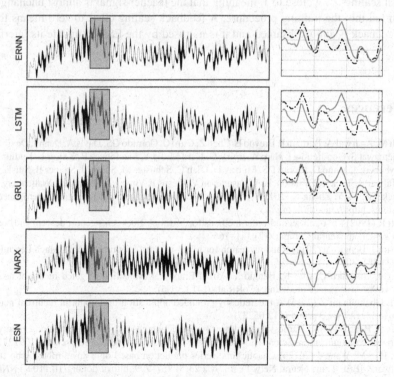

Fig. 7.5 GEFCom dataset—The plots on the left show the residuals of predictions of each RNN with respect to the ground truth; black areas indicate the errors in the predictions. The plots on right depict a magnification of the area in the gray boxes from the left graphics; the dashed black line is the ground truth, the solid gray line is the prediction of each RNN

The optimal ERNN configuration consists of only 60 nodes.

For LSTM, the optimal configuration includes only 20 hidden units, which is the lowest amount admitted in the validation search and SGD is the best as optimizer.

The optimal configuration for GRU is characterized by a large BPTT window, which assumes the maximum value allowed. This means that the network benefits from considering a large amount of past values to compute the prediction. As in LSTM, the number of processing units is very low. The best optimizer is Adam initialized with a particularly small learning rate, which yields a slower but more precise gradient update.

The optimal configuration of NARX network is characterized by a quite large number of hidden nodes and layers, which denote a network of higher complexity with respect to the ones identified in the other tasks. This can be related to the TDL larger values, which require to be processed by a network with greater modeling capabilities.

For ESN, we notice an extremely large spectral radius, close to the maximum value admitted in the random search. Consequently, also the value of the input scaling is set to a high number to increase the amount of nonlinearity in the processing units. The output scaling is set close to 1, meaning that the teacher signal is almost unchanged when fed into the training procedure. A feedback scaling close to zero means that the feedback is almost disabled and it is not used by the ESN to update its internal state.

References

Abadi M, Agarwal A, Barham P, Brevdo E, Chen Z, Citro C, Corrado GS, Davis A, Dean J, Devin M, Ghemawat S, Goodfellow I, Harp A, Irving G, Isard M, Jia Y, Jozefowicz R, Kaiser L, Kudlur M, Levenberg J, Mané D, Moore DS, Murray D, Olah C, Schuster M, Shlens J, Steiner B, Sutskever I, Talwar K, Tucker P, Vanhoucke V, Vasudevan V, Viégas F, Vinyals O, Warden P, Wattenberg M, Wicke M, Yu Y, Zheng X (2015) TensorFlow: Large-scale machine learning on heterogeneous systems. http://tensorflow.org/

Battiti R (1992) First-and second-order methods for learning: Between steepest descent and Newton's method. Neural Comput 4(2):141–166

Bergstra J, Bengio Y (2012) Random search for hyper-parameter optimization. J Mach Learn Res 13(1):281–305

Chung J, Gülçehre Ç, Cho K, Bengio Y (2014) Empirical evaluation of gated recurrent neural networks on sequence modeling. (CoRR abs/1412.3555)

Gal Y, Ghahramani Z (2015) A theoretically grounded application of dropout in recurrent neural networks. (ArXiv e-prints 1512:05287)

Jaeger H, Haas H (2004) Harnessing nonlinearity: Predicting chaotic systems and saving energy in wireless communication. Science 304(5667):78–80. https://doi.org/10.1126/science.1091277

Li D, Han M, Wang J (2012b) Chaotic time series prediction based on a novel robust echo state network. IEEE Trans Neural Netw Learn Syst 23(5):787–799. https://doi.org/10.1109/TNNLS.2012.2188414

Løkse S, Bianchi FM, Jenssen R (2017) Training echo state networks with regularization through dimensionality reduction. Cogn Comput pp 1–15. https://doi.org/10.1007/s12559-017-9450-z

Pham V, Bluche T, Kermorvant C, Louradour J (2014b) Dropout improves recurrent neural networks for handwriting recognition. In: 2014 14th International Conference on Frontiers in Handwriting Recognition. pp 285–290. https://doi.org/10.1109/ICFHR.2014.55

Shi Z, Han M (2007) Support vector echo-state machine for chaotic time-series prediction. IEEE Trans Neural Netw 18(2):359–372

Srivastava N, Hinton G, Krizhevsky A, Sutskever I, Salakhutdinov R (2014) Dropout: A simple way to prevent neural networks from overfitting. J Mach Learn Res 15(1):1929–1958

Theano Development Team (2016) Theano: A Python framework for fast computation of mathematical expressions. (arXiv e-prints abs/1605.02688)

Williams RJ, Zipser D (1995) Gradient-based learning algorithms for recurrent networks and their computational complexity. In; Backpropagation: Theory, architectures, and applications, vol 1. pp 433–486

Zeyer A, Doetsch P, Voigtlaender P, Schlüter R, Ney H (2016) A comprehensive study of deep bidirectional LSTM rnns for acoustic modeling in speech recognition. (CoRR abs/1606.06871)

Chapter 8
Conclusions

Abstract In this chapter we summarize the main points of our overview and draw our conclusions. We discuss our interpretations about the reasons behind the different results and performance achieved by the Recurrent Neural Network architectures analyzed. We conclude by hypothesizing possible guidlines for selecting suitable models depending on the specific task at hand.

Keywords Recurrent neural networks · Deep learning · Time series analysis · Short term load forecasting

In this overview, we investigated the application of recurrent neural networks to time series prediction, focusing on the problem of short-term load forecasting. We reviewed five different architectures, ERNN, LSTM, GRU, NARX, and ESN, explaining their internal mechanisms, discussing their properties and the procedures for the training. We performed a comparative analysis of the prediction performance obtained by the different networks on several time series, considering both synthetic benchmarks and real-world short-term forecast problems. For each network, we outlined the scheme we followed for the optimization of its hyperparameters. Relative to the real-world problems, we discussed how to preprocess the data according to a detailed analysis of the time series. We completed our analysis by comparing the performance of the RNNs on each task and discussing their optimal configurations.

From our experiments, we can draw the following important conclusions.

There is not a specific RNN model that outperforms the others in every prediction problem. The choice of the most suitable architecture depends on the specific task at hand and it is important to consider more training strategies and configurations for each RNN. On average, the NARX network achieved the lowest performance, especially on synthetic problems NARMA and MSO, and on the GEFCom dataset.

The training of gradient-based networks (ERNN, LSTM, and GRU) is slower and in general more complex, due to the unfolding and backpropagation through time procedure. However, while some precautions need to be taken in the design of these networks, satisfactory results can be obtained with minimal fine-tuning and by

F. M. Bianchi et al., *Recurrent Neural Networks for Short-Term Load Forecasting*,
SpringerBriefs in Computer Science, https://doi.org/10.1007/978-3-319-70338-1_8

selecting default hyperparameters. This implies that a strong expertise on the data domain is not always necessary.

The results obtained by the ESN are competitive in most tasks and the simplicity of its implementation makes it an appealing instrument for time series prediction. ESN is characterized by a faster training procedure, but the performance heavily depends on the hyperparameters. Therefore, to identify the optimal configuration in the validation phase, ESN requires a search procedure of the hyperparameters that is more accurate than in gradient-based models.

Another important aspect highlighted by our results is that the gated RNNs (LSTM and GRU) did not perform particularly better than an ERNN, whose architecture is much simpler, as well as its training. While LSTM and GRU achieve outstanding results in many sequence learning problems, the additional complexity of the complicated gated mechanisms seems to be unnecessary in many time series predictions tasks.

We hypothesize as a possible explanation that in sequence learning problems, such as the ones of Natural Language Processing (Kumar et al. 2016), the temporal dependencies are more irregular than in the dynamical systems underlying the load time series. In natural language, for example, the dependency from a past word can persist for a long time period and then terminate abruptly when a sentence ends. Moreover, there could exist relations between very localized chunks of the sequence. In this case, the RNN should focus on a specific temporal segment.

LSTM and GRU can efficiently model these highly nonlinear statistical dependencies, since their gating mechanisms allow to quickly modify the memory content of the cells and the internal dynamics. On the other hand, traditional RNNs implement smoother transfer functions and they would require a much larger complexity (number of units) to approximate such nonlinearities. However, in dynamical systems with dependencies that decay smoothly over time, the features of the gates may not be necessary and a simple RNN could be more suitable for the task.

Therefore, we conclude by arguing that ERNN and ESN may represent the most convenient choice in time series prediction problems, both in terms of performance and simplicity of their implementation and training.

Reference

Kumar A, Irsoy O, Ondruska P, Iyyer M, Bradbury J, Gulrajani I, Zhong V, Paulus R, Socher R (2016) Ask me anything: Dynamic memory networks for natural language processing. In: Balcan MF, Weinberger KQ (eds) Proceedings of the 33rd International Conference on Machine Learning, PMLR, New York, USA, Proceedings of Machine Learning Research, vol 48, pp 1378–1387

Printed in the United States
By Bookmasters